TESSELLATIONS

TESSELLATIONS
Mathematics, Art, and Recreation

Robert Fathauer

CRC Press
Taylor & Francis Group
Boca Raton London New York

CRC Press is an imprint of the
Taylor & Francis Group, an **informa** business

AN A K PETERS BOOK

First edition published 2021
by CRC Press
6000 Broken Sound Parkway NW, Suite 300, Boca Raton, FL 33487-2742

and by CRC Press
2 Park Square, Milton Park, Abingdon, Oxon, OX14 4RN

© 2021 Taylor & Francis Group, LLC

CRC Press is an imprint of Taylor & Francis Group, LLC

Library of Congress Cataloging-in-Publication Data
Names: Fathauer, Robert W., author.
Title: Tessellations : mathematics, art, and recreation / Robert Fathauer.
Description: Boca Raton : AK Peters/CRC Press, 2021. | Includes
bibliographical references and index.
Identifiers: LCCN 2020020936 (print) | LCCN 2020020937 (ebook) | ISBN
9780367185961 (paperback) | ISBN 9780367185978 (hardback) | ISBN
9780429197123 (ebook)
Subjects: LCSH: Tessellations (Mathematics)
Classification: LCC QA166.8 .F384 2020 (print) | LCC QA166.8 (ebook) |
DDC 516/.132--dc23
LC record available at https://lccn.loc.gov/2020020936
LC ebook record available at https://lccn.loc.gov/2020020937

ISBN: 9780367185961 (pbk)
ISBN: 9780367185978 (hbk)
ISBN: 9780429197123 (ebk)

Typeset in Avenir
by Deanta Global Publishing Services, Chennai, India

Contents

About the Author

Robert Fathauer has had a life-long interest in art but studied physics and mathematics in college, going on to earn a PhD from Cornell University in electrical engineering. For several years he was a researcher at the Jet Propulsion Laboratory in Pasadena, California. Long a fan of M.C. Escher, he began designing his own tessellations with lifelike motifs in the late 1980s. In 1993, he founded a business, Tessellations, to produce puzzles based on his designs. Over time, Tessellations has grown to include mathematics manipulatives, polyhedral dice, and books.

Dr. Fathauer's mathematical art has always been coupled with recreational math explorations. These include Escheresque tessellations, fractal tilings, and iterated knots. After many years of creating two-dimensional art, he has recently been building ceramic sculptures inspired by both mathematics and biological forms. Another interest of his is photographing mathematics in natural and synthetic objects, particularly tessellations. In addition to creating mathematical art, he's strongly committed to promoting it through group exhibitions at both the Bridges Conference and the Joint Mathematics Meetings.

Preface

This book is a synthesis of my "tessellations journey", which began in the late 1980s when I started trying to create Escheresque tessellations of my own. Two books were indispensable in my early explorations, *Tilings and Patterns*, by Branko Grünbaum and G.C. Shephard (1987), and *Visions of Symmetry: Notebooks, Periodic Drawings and Related Work of M.C. Escher*, by Doris Schattschneider (1990). The former is the widely acknowledged bible on the topic, and the latter contains reproductions of all of M.C. Escher's periodic drawings, along with in-depth analysis.

Other people who have encouraged and influenced me along the way include Chaim Goodman-Strauss, Edmund Harriss, Craig Kaplan, and Doris Schattschneider. Conferences with like-minded individuals have been a major source of inspiration as well as an outlet for sharing my work with others. My first mathematics/art conference was Nat Friedman's gathering in Albany, New York in the summer of 1997. This was followed by the Escher Centennial Congress in Rome and Ravello, Italy in 1998, and in 1999 by the first of 20 Bridges Conferences I've attended. Bridges, about connections between mathematics and the arts, was begun and made into the exciting and dynamic conference it is today by the tireless efforts of Reza Sarhangi.

Tilings and Patterns is difficult reading for a non-specialist, but its comprehensive descriptions and figures showing different classes of tilings was an invaluable reference in the age before Wikipedia and other websites on tiling. A lot of the content I found most helpful in *Tilings and Patterns* is included in this book, with two key differences. This book is written to be accessible to the non-specialist (i.e., people who are not professional mathematicians). In addition, this book is in full color compared to the black-and-white print of *Tilings and Patterns*, allowing the rich beauty of tessellations to be displayed.

This book is roughly divided into three main sections. The first ten chapters, not including Chapter 9, cover the topic of tiling in considerable depth. This is largely the content one would expect in a book on the topic. Chapters 11–17 explore the creation of Escheresque planar tessellations, and Chapters 20–25 treat the application of Escheresque tessellations to polyhedra and other surfaces. Interspersed are chapters treating special topics that I've delved into over the years: fractal tiling (Chapter 9), decorating tiles with knots and other graphics (Chapter 18), and tessellation metamorphoses (Chapter 19).

Exploring the art and recreational mathematics aspects of tessellations was my main preoccupation for many years. I spent several thousand hours blissfully lost in such explorations, particularly searching for and discovering new fractal tilings. Capturing the beauty of tessellations and conveying it to others is a pastime that never grows old for me. Unless otherwise noted, all the graphics and photographs in this book are my own work.

It would be hard to name another topic with such strong visual appeal, so many practical applications, and so much entertaining mathematics. It's my hope that this book will spark the same love (if not obsession!) that I've experienced with this topic in those who read it – especially in students, who too are often inadequately exposed to the beauty and fun of mathematics.

Chapter 1

Introduction to Tessellations

A **tessellation** is a collection of shapes that fit together without gaps or overlap to cover the infinite mathematical plane. Another word for tessellation is **tiling**, and the individual shapes in a tessellation are referred to as **tiles**. In a more general sense, a tessellation covers any surface, not necessarily flat or infinite in extent, without gaps or overlaps.

A tessellation is a type of **pattern**, which can be defined as a two-dimensional design that (generally) possesses some sort of symmetry. Symmetries in one, two, and three dimensions have been classified mathematically, and these classifications apply to all types of patterns, including tessellations. Symmetry and transformation in tessellations are the topics of Chapter 3.

Tilings in which the individual tiles are recognizable, real-world motifs will often be referred to as **Escheresque tessellations**

in this book (after Dutch graphic artist M.C. Escher), and the tiles in them as Escheresque tiles. These sorts of designs are a relatively recent phenomenon, dating back a little over a century. Tilings in which the individual tiles are geometric shapes rather than recognizable motifs will often be referred to as **geometric tessellations** in this book. Types and properties of geometric tessellations will be explored in the next chapter, and some particular types of geometric tessellations will be treated in greater depth in Chapters 6–10. Chapters 11–17 are devoted to planar Escheresque tessellations. Chapters 18 and 19 deal with specialized topics, and Chapters 20–25 deal with applying Escheresque tessellations to three-dimensional surfaces. The terms "tiling" and "tessellation" will be used interchangeably in this book.

Historical examples of tessellations

The use of patterns by people is ubiquitous and must surely predate recorded history in virtually every society. There is a basic human impulse to decorate, personalize, and beautify our surroundings by applying patterns to textiles, baskets, pavers, tiles, and even our own bodies. As a result, tessellations are found in virtually all cultures from very early on. Tessellations have been and continue to be used in tribal art from around the world for decoration of baskets (Figure 1.1), blankets, masks, and other objects.

The word "tessellation" comes from the Greek word *tessares*, meaning four, and the Latin *tessellare*, meaning to pave with tesserae. Tessera (plural form tesserae) are small squares or cubes of stone or glass used for creating mosaics. The Ancient Romans often used tesserae to form larger geometric figures, as well as figures such as animals or people (Figure 1.2). In addition, they used

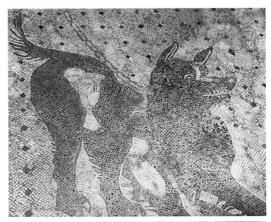

Figure 1.2. Roman mosaics and tilings in Pompeii, Italy.

geometric tiles like rhombi (sometimes called diamonds) to form tessellations as decorative floor coverings.

Beautiful examples of tessellations and related patterns in architecture dating back several centuries can be found in monumental

Figure 1.1. Contemporary African basket in which tessellated regions are used for decoration.

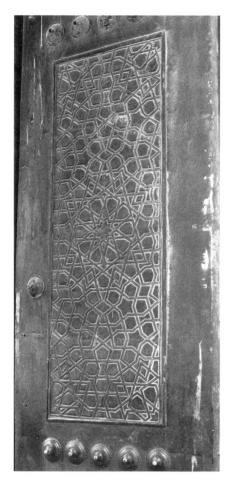

Figure 1.3. Door in the Sultan Ahmed (Blue) Mosque, Istanbul, Turkey.

Figure 1.4. Mosaic from the floor of Saint-Sernin Basilica, a Romanesque building in Toulouse, France.

Figure 1.5. Panel on the quarters of the "Dragon Lady" in the Forbidden City, Beijing, China.

buildings such as cathedrals and palaces around the world. These often have patterns and styles characteristic of particular cultures. Islamic art has an especially rich tradition in the use of distinctive and sophisticated geometric tessellations, starting around the 9th century and continuing to the present. Examples from the Middle East, Europe, and China are shown in Figures 1.3, 1.4, and 1.5, respectively.

During the Renaissance, artists and mathematicians alike were fascinated by polyhedra

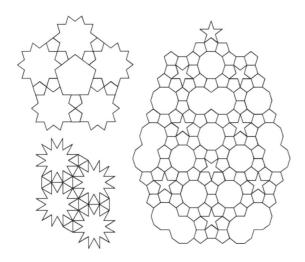

Figure 1.6. Tessellations by Johannes Kepler incorporating regular and star polygons (redrawn).

and tessellations. Albrecht Dürer (1471–1528), a painter and printmaker, designed a tessellation, incorporating pentagons and rhombi [Jardine 2018]. Johannes Kepler (1571–1630), most famous for discovering the laws of planetary motion, designed tessellations of regular and star polygons (Figure 1.6) that are included in his 1619 book *Harmonices Mundi* [Grünbaum 1987].

Tessellations in the world around us

A tessellation is a mathematical construction in which the individual tiles meet along a line of zero width. In any real-world tessellation, there will be some finite width to that line. In addition, no real-world plane is perfectly flat. Nor is any real shape perfectly geometric; e.g., no brick is a perfect rectangular solid. Finally, any real-world tessellation will obviously have a finite extent, while a mathematical tessellation extends to infinity. When we talk about tessellations in the real world, then, it should always be kept in mind that we are talking about something that does not strictly meet the definition of a tessellation given above. Another way of looking at it is that a tessellation that we associate with a physical object is a mathematical model that describes and approximates something in the real world. Many tessellations in nature are highly irregular, as seen in the example of Figure 1.7. A variety of other examples will be presented in Chapter 4.

Tessellations have been widely used in architecture, not only purely for decoration but also by necessity from the nature of building materials like bricks (Figure 1.8). These same types of uses continue today and are so widespread that one is likely to find numerous examples on a short walk around most modern cities. Tessellations are also widely used in fiber arts, as well as games and puzzles. These sorts of uses are explored in more detail in Chapter 5.

Figure 1.7. A tessellation created by the cracking of a lava flow in Hawaii. Many of the individual tiles can be approximated as pentagons and hexagons.

Figure 1.8. Tessellated brickwork on the steeple of a church in the Friesland region of the Netherlands.

Escheresque tessellations

Lifelike motifs were incorporated early on in decorative patterns and designs, and some of these might be considered tessellations. Koloman Moser, an Austrian designer and artist who lived from 1868 to 1918, is generally credited with creating the first tessellations, in which the individual tiles depict recognizable, real-world objects. In addition to painting, Moser designed jewelry, glass, ceramics, furniture, and more. As a leading member of the Vienna Secession movement, he was dedicated to fine craftsmanship, and his work incorporated stylized versions of forms found in nature.

The tessellation shown in Figure 1.9 was published in 1901–1902 as part of a work

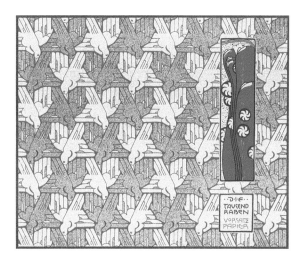

Figure 1.9. Tessellation of ravens by Koloman Moser, dating to 1902. Imaging Department © President and Fellows of Harvard College.

titled *Flächenschmuck* (planar ornamentation), volume 3 of the series Die Quelle. This work contains other designs that can be considered Escheresque tessellations.

M.C. Escher is the name most often associated with tessellations in which the tiles are real-world motifs. Escher lived from 1898 to 1972. Early in his career, his work depicts landscapes and other conventional themes. He became fascinated with tiling when he visited the Alhambra, a Moorish palace in Granada, Spain, in 1922. He made drawings of several of the geometric tilings he saw there. His first tessellations, in which the tiles were lifelike motifs, were made in 1926–1927, and in 1936, he began producing a large number of such designs. He kept notebooks in which he enumerated 137 of these designs and recorded his own classification system for them. His notebooks are meticulously reproduced and analyzed in Doris Schattschneider's book *Visions of Symmetry – Notebooks, Periodic Drawings, and Related Work of M. C. Escher* [Schattschneider 2004]. He remained fascinated by these sorts of designs for the remainder of his life, with his last numbered design being produced in 1971. In Chapters 13–17, which contain tessellation templates, there are several notes on designs of Escher's that can be created with the templates.

Escher incorporated many of his tessellations into woodcuts and lithographs that became well known. Some of the most popular of these involve the metamorphosis of a tessellation from geometric tiles to real-world motifs. Tessellation metamorphoses are the subject of Chapter 19. Other Escher prints play on the two-dimensional nature of tessellations and prints vs. the three-dimensional nature of our world. For example, in his 1943 lithograph "Reptiles" (Figure 1.10), a tessellation of reptiles drawn on a paper is shown laying on a table. At one edge, the reptiles become three-dimensional, crawl out of the paper, make a circuit, and crawl back in on the other side.

Escher's tessellations have inspired many mathematicians and artists. Starting roughly around 1990, several artists put considerable time and effort into designing their own tessellations in which the tiles are real-world motifs. Prior to that time, there were only scattered attempts at such designs. These artists include Yoshiaki Araki [Araki 2020], Bruce Bilney [Bilney 2020], David Bailey, Seth Bareiss [Bareiss 2020], Andrew Crompton (example in Figure 1.11) [Crompton 2020], Hollister David [David 2020], Robert Fathauer, Hans Kuiper [Kuiper 2020], Ken Landry [Landry 2020], Makoto Nakamura [Nakamura 2020], John Osborn [Osborn 2020], Peter Raedschelders [Raedschelders 2020], Marjorie Rice [Rice 2020], and Patrick Snels. Patrick has created a database containing the work of numerous

Figure 1.10. Escher's lithograph "Reptiles" (1943), which incorporates his 25th periodic drawing. M.C. Escher's "Reptiles" © 2020 The M.C. Escher Company, The Netherlands. All rights reserved. www.mcescher.com.

Figure 1.11. "Cheshire Cats", a digital spherical tessellation by Andrew Crompton in 2000. Image courtesy of Andrew Crompton.

tessellation artists [Snels 2019]. Examples of the work of these and other artists will be sprinkled throughout the following chapters. Any Escheresque tessellation in this book without the artist named in the caption is my own work.

Some of these artists have pushed tessellations in new directions not explored by Escher. John Osborn and Robert Fathauer have created Escheresque tiles that can be assembled in many different ways, following Sir Roger Penrose's creation of a hen design for his famous aperiodic tiles. These designs have been used as the basis for mass-produced puzzles. Peter Raedschelders, Robert Fathauer, Richard Hassell, and Alain Nicolas have created fractal tessellations using Escheresque tiles (Figure 1.12). Fractals are a branch of mathematics that was only brought to light near the end of Escher's life. Chapter 9 is devoted to fractal tilings. Ken Landry has applied his tessellations to the surfaces of complex polyhedra, following similar application of Escher's designs in the 1977 book *M.C. Escher Kaleidocycles*, by Doris Schattschneider and Wallace Walker [Schattschneider 1987]. Chapters 21–24 deal with applying tessellations to polyhedra.

Figure 1.12. "Bats and Owls", a fractal Escheresque tessellation screen print by Robert Fathauer from 1993.

Makoto Nakamura has combined multiple tessellations in single scenes that are colorful and full of movement. Douglas Dunham has used computer programming to create hyperbolic tessellations using Escher's designs that have different symmetries from those Escher created. Dominique Ribault and others have created tessellations on spheres and tori. Hyperbolic and spherical tessellations are explored in more depth in Chapter 9. Bruce Bilney and Patrick Snels have created animations of kangaroos and reptiles jumping and crawling that are tessellated in every frame. Makoto Nakamura and Alexander Guerten (Figure 1.13) have made three-dimensional figures that tessellate space.

Figure 1.13. 3D-printed tessellated animals created by Alexander Guerten in 2018. Copyright 2017 Alexander Guerten, contralex.com.

Tessellations and recreational mathematics

There have been many games and puzzles based on geometric and Escheresque tessellations. Jigsaw puzzles can also be thought of as tessellation puzzles, in which the tiles have imagery printed on them. In addition, any grid can be thought of as a tessellation, and grids are ubiquitous in puzzles and games, from tic-tac-toe to chess to sudoku.

Apart from traditional recreational activities, many people enjoy solving math problems and playing with mathematical objects as a pastime. Figuring out how to create an Escheresque tessellation using a particular motif or geometric basis is mathematical recreation as well as art.

Polyominoes, "invented" in the 1950s by Solomon Golomb, have been a rich source of mathematical recreation [Golomb 1994]. *A **polyomino** is a shape made up of squares attached in an edge-to-edge fashion.* There

are five different tetrominoes, shapes made up of four squares. These are the basis for the computer game Tetris, designed by the Russian computer engineer Alexey Pajitnov in 1985. In Tetris, falling tetrominoes have to be oriented and positioned to fit into a growing tessellation. In schools, pentominoes have been widely used for teaching about math and problem-solving. These are comprised of 5 squares, and there are 12 distinct pentominoes. Analogous families of

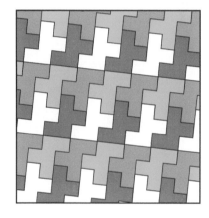

Figure 1.14. A tessellation of multiple copies of the same pentomino.

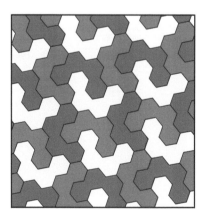

Figure 1.15. A tessellation of multiple copies of the same polyhex.

shapes based on other polygons have been invented as well. **Polyhexes** *are shapes made up of regular hexagons, and* **polyiamonds** *are shapes made up of equilateral triangles.* Examples of pentomino and polyhex tilings are shown in Figures 1.14 and 1.15. These various sets of tiles are explored in more depth in Chapter 6.

In recent years, a sizable community has been developed of people who combine tessellation and origami, including Alessandro Beber (example in Figure 1.16), Eric Gjerde [Gjerde 2009], Robert Lang, Ekaterina Lukasheva, and Chris Palmer. This technique usually entails creating a dense grid of creases in a piece of paper followed by folding to

Figure 1.16. An example of tessellation origami, entitled "R-3.4.6.4", created by Alessandro Beber in 2012. Copyright 2012 Alessandro Beber.

generate tessellations of regular polygons, star polygons, etc.

Tessellations are also a topic studied by mathematicians. One of the most famous mathematical discoveries in this field is the set of two aperiodic tiles discovered by English mathematician and physicist Roger Penrose in 1973/1974. One version of these, known as "kites and darts", is shown in Figure 1.17. The tiles have matching rules that prevent periodic arrangements. In 1982, Dan Shechtman observed diffraction patterns from an alloy that exhibited mathematically forbidden ten-fold symmetry. He received the Nobel Prize in chemistry for this work in 2011. Penrose tiles were used to help explain this result. Aperiodic tilings are examined in more depth in Chapter 8.

In 2015 a new distinct tiling of the plane by convex pentagons, discovered by Casey Mann, Jennifer McLoud, and David Von Derau, and shown in Figure 1.18, got considerable attention from the press [Mann 2015]. The discovery was made using computer search methods, which have been applied to other tiling problems including polyominoes. Many examples of tessellations by pentagons and other polygons are presented in the next chapter.

Figure 1.17. A portion of a tiling of Penrose kites and darts. No infinite tiling of this sort will be periodic or repeating.

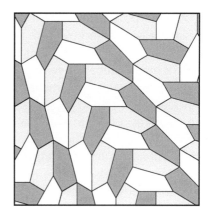

Figure 1.18. Convex pentagon tiling discovered in 2015 by Casey Mann et al.

Tessellations and mathematics education

Over the course of the 1990s, tessellations became widely taught in schools in the United States. There were several contributing factors behind this. One was that teachers became aware of the educational value of teaching tessellations, particularly for the math content (see list below). At the same time, teachers realized that students love Escheresque tessellations. When students are excited about an activity, they are more engaged and focused. In parallel, books and software appeared on the market that facilitated the understanding and use of tessellations. The 1989 book *Introduction to Tessellations*, by Dale Seymour and Jill Britton, helped raise awareness of tessellations and make the topic more accessible [Seymour 1989]. M.C. Escher's tessellations were published in full for the first time and analyzed in the 1990 book *Visions of Symmetry – Notebooks, Periodic Drawings, and Related Work of M.C. Escher*, by Doris Schattschneider [Schattschneider 2004]. A 1994 computer program called *TesselMania!*, written by Kevin Lee, made it easier for students to design their own tessellations. In 1993, my own company, called Tessellations, produced its first tessellation puzzle, *Squids & Rays*, which was enthusiastically received by math teachers. Unfortunately, the appreciation of tessellations in K–12 math education seems to have diminished in recent years. A number of factors likely contributed to this, and one of the goals of this book is to rekindle that excitement in a new generation of math teachers and students. Classroom activities are included at the end of every chapter in this book.

In 2010, the Common Core State Standards for Mathematical (CCSSM) Practice were published in the United States. These standards have been adopted by a majority of states, and many states require classroom materials to be aligned with the standards. While tessellations and polyhedra are not topics that are explicitly addressed in the CCSSM, a number of the standards are addressed by the general content of this book and in the activities. Some generally relevant standards are quoted below, and many of the activities quote specific standards.

- Recognize a line of symmetry for a two-dimensional figure. (4.G 3)
- Identify line-symmetric figures and draw lines of symmetry. (4.G 3)
- Recognize angle measure as additive. (4.MD 7)
- Recognize angles as geometric shapes that are formed wherever two rays share a common endpoint and understand concepts of angle measurement. (4.MD 5)
- Represent three-dimensional figures using nets made up of rectangles and triangles, and use the nets to find the surface area of these figures. (6.G 4)
- Describe the two-dimensional figures that result from slicing three-dimensional figures, as in plane sections of right rectangular prisms and right rectangular pyramids. (7.G3)

- Solve real-world and mathematical problems involving area, volume, and surface area of two- and three-dimensional objects composed of triangles, quadrilaterals, polygons, cubes, and right prisms. (7.G 6)
- Understand that a two-dimensional figure is congruent to another if the second can be obtained from the first by a sequence of rotations, reflections, and translations (8.G 2)
- Apply the Pythagorean Theorem to determine unknown side lengths in right triangles in real-world and mathematical problems in two and three dimensions. (8.G 7)
- Know the formulas for the volumes of cones, cylinders, and spheres and use them to solve real-world and mathematical problems. (8.G.9)
- Rearrange formulas to highlight a quantity of interest, using the same reasoning as in solving equations. For example, rearrange Ohm's law $V = IR$ to highlight resistance R. (A-CED 4)
- Experiment with transformations in the plane. (G-CO)
- Given a rectangle, parallelogram, trapezoid, or regular polygon, describe the rotations and reflections that carry it onto itself. (G-CO 3)
- Visualize relationships between two-dimensional and three-dimensional objects. (G-GMD)
- Use volume formulas for cylinders, pyramids, cones, and spheres to solve problems. (G-GMD 3)
- Use geometric shapes, their measures, and their properties to describe objects. (G-MG 1)

For many of the activities in this book, a specific standard addressed is spelled out in the activity instructions.

Activity 1.1. Recognizing tessellations

Throughout the book, the activities are worded for teachers working with students. However, the activities can also be carried out by individuals who can answer the discussion questions after completing each activity.

> *Materials:* Copies of Worksheet 1.1.
> *Objective:* Learn to tell if a collection of shapes is a tessellation or not.
> *Vocabulary:* Tessellation, infinite, mathematical plane.

Activity Sequence
1. Write the definition of a tessellation on the board. Be sure the class understands it and also what the terms infinite and mathematical plane mean.
2. Pass out copies of the worksheet.
3. Have the students circle those patterns that are tessellations and put an "X" through those that aren't.

Discussion Questions
1. Is pattern letter a/b/c/d/e/f a tessellation?
2. Why or why not?
3. If it isn't, how might it be changed to make it into a tessellation?

Worksheet 1.1. Recognizing tessellations

Circle the collections of gray squares that are tessellations; mark "X" through those that aren't.

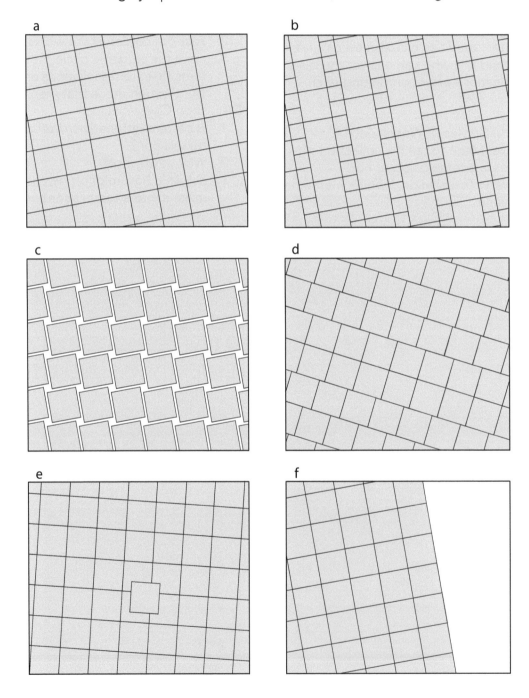

a

b

c

d

e

f

Activity 1.2. Historical tessellations

Materials: Copies of the worksheet.

Objective: Learn about the variety of tessellations from different time periods and cultures.

Vocabulary: Islamic, star polygon, tessera, motif, Middle Ages, rhombus.

Activity Sequence

1. Write the vocabulary terms on the board and discuss the meaning of each.
2. Pass out copies of Worksheet 1.2.
3. Have the students draw lines connecting the word descriptions to the representative tessellations.

Discussion Questions

1. Which tessellation corresponds to the first-word description? How do you know?
2. Which tessellation corresponds to the second-word description? How do you know?
3. Which tessellation corresponds to the third-word description? How do you know?
4. Which tessellation corresponds to the fourth-word description? How do you know?
5. Which tessellation corresponds to the fifth-word description? How do you know?
6. Which tessellation corresponds to the sixth-word description? How do you know?

Worksheet 1.2. Historical tessellations

Draw lines to match the paragraph describing the use of tessellations in various times and cultures in the left with the representative examples in the right.

1. Islamic architecture features complex geometric tessellations that often incorporate star polygons.

2. Ancient Roman floor tilings utilized small square stones called tesserae, with which larger geometric patterns and figures were formed.

3. Bricks have been widely used to pave sidewalks and roads.

4. Around the year 1900, Koloman Moser designed tessellations in which the individual tiles are recognizable, real-world motifs.

5. In the Middle Ages, before large plate glass technology existed, windows were made up of a number of small, square, rectangular, or rhombus-shaped panes.

6. "Tetris" is a computer game in which falling tetrominoes must be fit together. Tetrominoes are shapes made by connecting four squares.

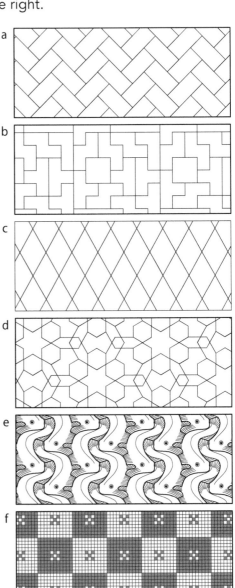

a

b

c

d

e

f

Chapter 2

Geometric Tessellations

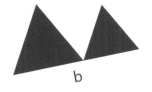

a

b

In order to explore tessellations more fully, it will be helpful to have a better understanding of the mathematics behind them. This chapter introduces the terminology of tessellations and uses that terminology to describe many important concepts related to tessellations. Definitions of terms are italicized, with the term itself bolded. Don't worry if you can't remember all these terms or don't fully understand some of them. You can still use the later chapters in this book and enjoy tessellations. As you become more familiar with tessellations, you'll come to appreciate the use of more precise terminology and more advanced concepts.

Tiles

Recall that a tessellation is a collection of shapes, called tiles, that fit together without gaps or overlaps to cover the infinite plane.

Let's look more closely at tiles. A mathematical definition of a tile is a closed topological disk. In simpler terms, *a **tile** is a set whose boundary is a single simple closed curve.* In this context, the term curve includes straight-line segments, so basically any single two-dimensional shape without holes is a tile. Tiles also shouldn't neck down to points or lines anywhere. Some examples of things that are not tiles are shown in Figure 2.1. *The area inside the curve, or boundary, is the **interior** of the tile.*

A ***polygon*** *is a closed plane figure made up of straight-line segments.* For a polygonal tile, *the individual segments are referred to as **edges**, and the points where two edges meet are referred to as **corners*** (Figure 2.1). Examples of tiles that are not polygonal are shown in Figure 2.3.

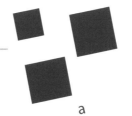

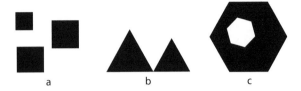

Figure 2.1. Three examples of things that are not tiles. (a) A collection of unconnected shapes; (b) two shapes touching at a point; (c) a shape with a hole in it.

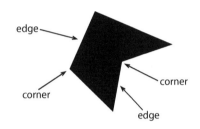

Figure 2.2. A polygonal tile has a boundary consisting of edges and corners.

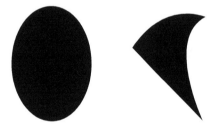

Figure 2.3. Besides being polygons, tiles can have no straight lines or a mixture of straight and curved lines.

Angles

For a polygonal tile, *the angle between two adjacent edges inside the tile is called the* **interior angle**. *The angle between the same two edges outside the tile is called the* **exterior angle**. A full revolution measures 360° (Figure 2.4). As a result, for any two adjacent edges of a tile, the interior and exterior angles always sum to 360°. Some common angles encountered in tessellations, particularly those of regular polygons, are shown in Figure 2.5. Each angle is specified in three different ways: in degrees, as a fraction of π, and as a fraction of a full revolution. A full revolution equals 360° or 2π. These comments about angles pertain to the **Euclidean plane**, *the two-dimensional* geometric space students learn about in K–12 education. Unless otherwise stated, tessellations in this book should be understood to live in this space.

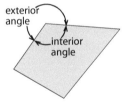

Figure 2.4. The interior and exterior angles between two adjacent edges sum to 360°.

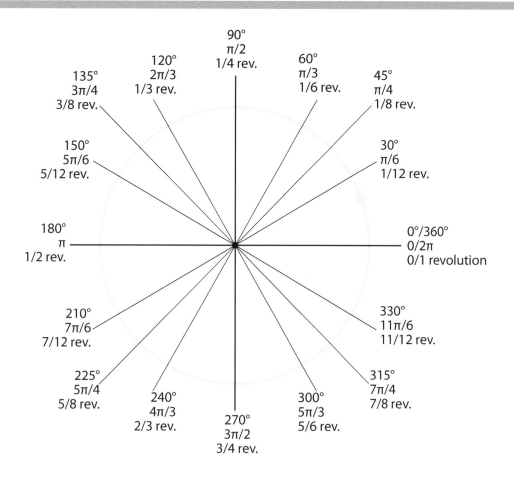

Figure 2.5. Common angles in tessellations.

Vertices and edge-to-edge tessellations

A point at which three or more tiles meet is called a **vertex** *(Figure 2.6). The sums of the interior angles of the tiles meeting at a vertex must be 360°, which is a full revolution. The number of tiles meeting at a vertex is the* **valence** *of that vertex.* Other terms that mean the same thing as valence, borrowing from the language of graph theory, are *the* **order** *of a vertex and the* **degree** *of a vertex.*

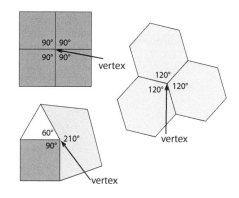

Figure 2.6. Examples of vertices.

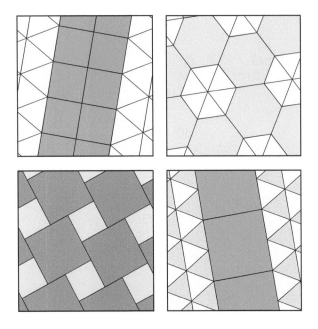

Figure 2.7. Tessellations that are edge-to-edge (top two) and not edge-to-edge (bottom two).

A tessellation of polygons is called **edge-to-edge** if adjacent tiles always touch along their entire edges. In other words, the corners of the tiles coincide with the vertices of the tessellation. In an edge-to-edge tiling, there cannot be a single vertex of the tiling that meets a single tile along an edge rather than at a corner. Examples of tessellations that are and aren't edge-to-edge are shown in Figure 2.7.

Regular polygons and regular tessellations

A **regular polygon** is one for which each edge is of the same length and each interior angle has the same measure. Regular polygons without too many sides have common word names, but it's also acceptable to refer to any regular polygon with *n* sides as a regular *n*-gon. In addition, the word "regular" is often not explicitly used. For example, the term "hexagon" is often understood to mean a regular hexagon, depending on the context. The interior angles of a regular *n*-gon are $(n - 2)\pi/n$. An **equilateral polygon** is one for which each edge is of the same length. The interior angles are not necessarily the same, however, so equilateral polygons are not necessarily regular. Some regular polygons commonly encountered in geometric tessellations are shown in Figure 2.8.

A **regular tessellation** is an edge-to-edge tessellation by congruent regular polygons. There are exactly three regular tessellations, based on the equilateral triangle, square, and regular hexagon (Figure 2.9). Note that there is a close relationship between the equilateral triangle and hexagon tessellations and that a regular hexagon can be divided into six equilateral triangles. While the equilateral triangle and square tessellations can be thought of as

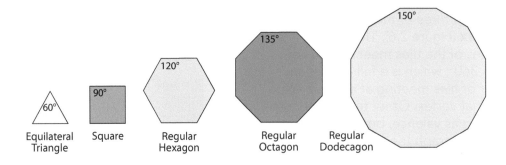

Figure 2.8. Some regular polygons that appear frequently in tessellations.

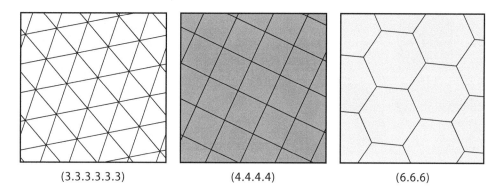

| (3.3.3.3.3.3) | (4.4.4.4) | (6.6.6) |

Figure 2.9. The three regular tessellations.

grids of parallel lines, this isn't the case for the hexagon tessellation. Finally, the equilateral triangles appear in two distinct orientations, while all the squares and hexagons appear in the same orientation.

Regular pentagons do not tessellate by themselves, as it is not possible for them to meet at a point without either leaving a gap or overlapping (Figure 2.10). Since equilateral triangles, squares, and regular hexagons form regular tessellations, that takes care of regular polygons through six sides. Any regular polygon with more than six sides has interior angles greater than 120°, and since 360°/3 = 120°, it's not possible for three or more of any of these regular polygons to meet at a

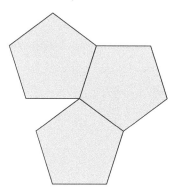

Figure 2.10. Regular pentagons do not tile the plane because it's not possible to form a vertex without gaps or overlaps.

point. Triangles, squares, and hexagons are therefore the only regular polygons that form regular tessellations.

Regular-polygon vertices

Vertices are characterized by the types of tiles that meet at them. If the same types of tiles meet at two vertices and in the same order, the vertices are considered to be of the same type. In the three regular tessellations, every vertex is of the same type.

In the case of edge-to-edge tessellations of regular polygons, vertices can be simply specified by the types of regular polygons that meet at them. Integers specify different regular polygons, e.g., an equilateral triangle is specified by the number 3 and a regular hexagon by the number 6. A vertex is then specified by listing the regular polygons meeting at it. By convention, the numbers are listed in cyclic order in the direction that leads to a smaller lexicographical sequence. This just means the digits are compared one by one until a digit in one sequence has a smaller value.

It can be shown that there are exactly 21 distinct types of vertices (Figure 2.11) that can be formed by arranging regular polygons around a point [Grünbaum 1987]. However,

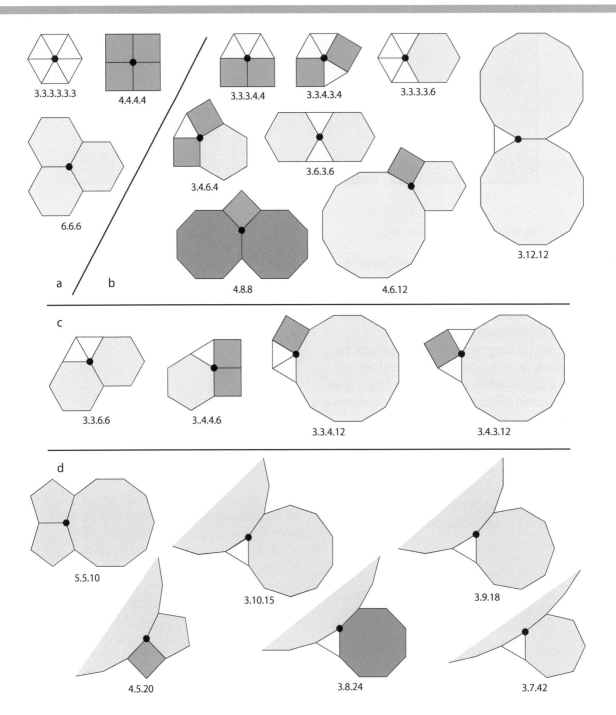

Figure 2.11. The 21 types of vertices that can be formed by regular polygons. The vertex types in area (a) generate the three regular tessellations. The vertex types in area (b) generate the eight semi-regular tessellations. There are edge-to-edge tilings of regular polygons that contain the vertices in area (c), but such tilings contain more than one type of vertex. There are no tilings of regular polygons that contain the vertices in area (d).

for six of these, there is no edge-to-edge tessellation by regular polygons that contains even a single vertex of that type. It is possible to form tessellations by combining rhombi with these regular-polygon vertices; a number of examples are explored in a website called "Imperfect Convergence" [Jardine 2018].

A group of tiles surrounding a vertex is an example of a patch. A **patch** *is a finite number of tiles whose union, like a single tile, is a topological disk.* Unlike tilings, patches do not cover the entire mathematical plane.

Prototiles

A tile to which many or all other tiles in a tessellation are similar (have the same shape) is called a **prototile**. The regular tessellations all have a single prototile, while the semi-regular tessellations described in the next section all have more than one prototile. *If a particular tessellation can be created using a particular prototile, that tile is said to* **admit** *the tessellation.* In this book, the term prototile refers only to the shape, i.e., a single-prototile tiling can have copies of the prototile of different sizes, and they may be mirrored.

A **monohedral** *tiling is one in which all tiles are congruent, and mirrors are allowed. The size as well as shape are the same in* **congruent** *tiles.* The regular tessellations are all monohedral. Note that regular polygons possess mirror symmetry, but this is not the case for tilings such as the general triangle tilings discussed later in this chapter. Similarly, *a* **dihedral** *tiling is one in which all tiles are congruent to one of the two prototiles.* Six of the semi-regular tessellations described in the next section are dihedral and two are trihedral (containing three distinct tiles).

Semi-regular tessellations

There are eight **semi-regular tessellations** *– edge-to-edge tessellations for which every tile is a regular polygon, each vertex is of the same type, and there is more than one type of regular polygon. The regular and semi-regular tessellations together are called the* **Archimedean tessellations**, *though this term is also used to mean only the semi-regular tessellations. The notation for these is the same as the notation for the vertices that generate them, with the addition of parentheses to signify a tiling as opposed to a vertex (Figure 2.12). One of the eight, the (3.3.3.3.6) tessellation, exists in two distinct forms that are mirror images of one another. The other seven possess reflection symmetry.*

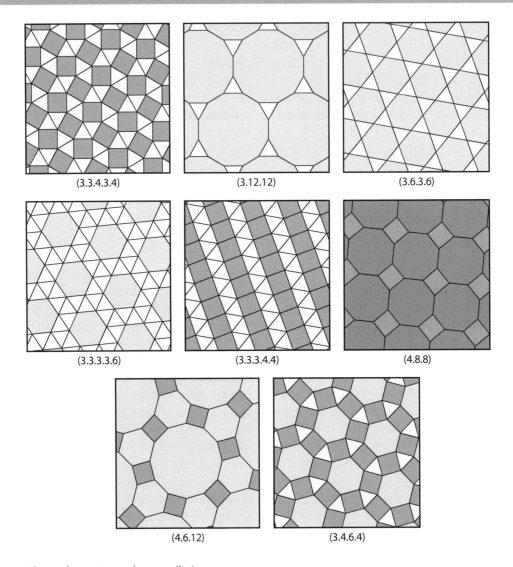

(3.3.4.3.4)   (3.12.12)   (3.6.3.6)

(3.3.3.3.6)   (3.3.3.4.4)   (4.8.8)

(4.6.12)   (3.4.6.4)

Figure 2.12. The eight semi-regular tessellations.

Other types of polygons

There are many types of polygons that commonly turn up in tessellations other than regular polygons. Some of these are named and described in Figure 2.13. Many of them will be referred to later in the book. Two that are not seen very commonly outside of tessellations are kites and darts, also known as convex kites and concave kites. *A **convex polygon** is one in which none of the interior angles exceed 180°, while a **concave polygon** is one in which at least one interior angle exceeds 180°.*

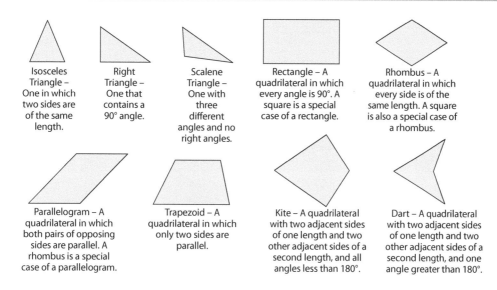

Figure 2.13. Several types of irregular polygons, with definitions.

Isosceles Triangle – One in which two sides are of the same length.

Right Triangle – One that contains a 90° angle.

Scalene Triangle – One with three different angles and no right angles.

Rectangle – A quadrilateral in which every angle is 90°. A square is a special case of a rectangle.

Rhombus – A quadrilateral in which every side is of the same length. A square is also a special case of a rhombus.

Parallelogram – A quadrilateral in which both pairs of opposing sides are parallel. A rhombus is a special case of a parallelogram.

Trapezoid – A quadrilateral in which only two sides are parallel.

Kite – A quadrilateral with two adjacent sides of one length and two other adjacent sides of a second length, and all angles less than 180°.

Dart – A quadrilateral with two adjacent sides of one length and two other adjacent sides of a second length, and one angle greater than 180°.

General triangle and quadrilateral tessellations

Earlier in the chapter, we saw that equilateral triangles and squares both tessellate. What about triangles that aren't equilateral and quadrilaterals other than squares? It turns out that any triangle and any quadrilateral will tessellate (Figure 2.14). Any triangle can be copied, rotated 180°, and joined in an edge-to-edge fashion to an unrotated triangle so that the two form a parallelogram. These parallelograms then tessellate in a straightforward manner. In the case of quadrilaterals, rows of tiles that alternate between rotated and unrotated versions form zigzag bands that fit together to cover the plane. There are many other tessellations of triangles and rectangles. A few examples are shown in Figure 2.15.

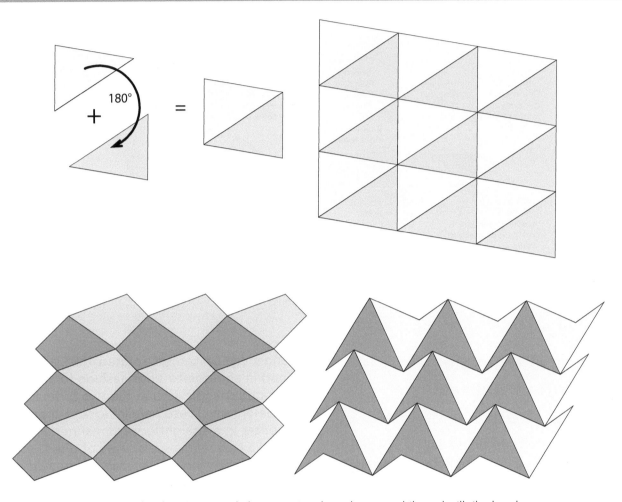

Figure 2.14. When paired with 180°-rotated tiles, any triangle and any quadrilateral will tile the plane.

Figure 2.15. Some other tessellations of triangles and quadrilaterals.

Dual tessellations and Laves tessellations

A **dual tessellation** is obtained by connecting the centers of every pair of adjacent tiles in a tessellation. Of particular interest are the **Laves tessellations** (Figure 2.16), which are duals of the Archimedean tessellations. The uniformity of the vertex locations in an Archimedean tessellation leads to all the tiles in a Laves tessellation being congruent. In three dimensions, an analogous situation occurs with Archimedean solids and their duals, the Catalan solids. All faces in a Catalan solid are identical. As a result of the tiles being congruent, the Laves tessellations can serve as templates for single-motif Escheresque tessellations. The Laves tessellations are also known as the Catalan tessellations.

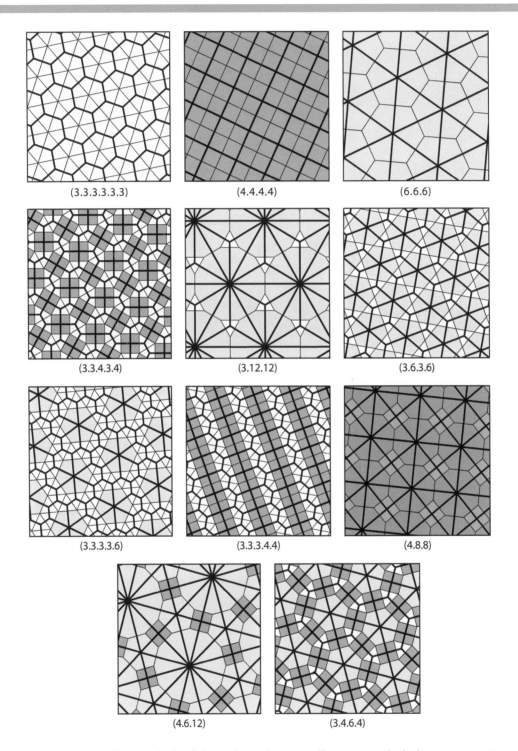

(3.3.3.3.3.3)   (4.4.4.4)   (6.6.6)

(3.3.4.3.4)   (3.12.12)   (3.6.3.6)

(3.3.3.3.6)   (3.3.3.4.4)   (4.8.8)

(4.6.12)   (3.4.6.4)

Figure 2.16. The Laves tessellations, duals of the Archimedean tessellations, on which they are superimposed in bold black lines. The descriptors pertain to the Archimedean tessellations.

Pentagon and hexagon tessellations

As shown earlier in the chapter, regular pentagons won't tessellate, but there are a variety of other pentagons that will tessellate. The exact number of distinct classes of tilings by congruent convex pentagons is a problem that has drawn considerable interest among mathematicians. The first five classes were described early in the 20th century by Karl Reinhardt [Wikipedia 2018], though at least some were known previously. Examples of three of these are found in the Laves tessellations, described in the preceding section. There was a gap of a half century before three more were discovered in 1968, followed by a ninth in 1975. Remarkably, another four were discovered over the next few years by a San Diego homemaker named Marjorie Rice, who developed her own notation to carry

out a systematic search after reading about the problem in an article by Martin Gardner [Schattschneider 1978]. A 14th was discovered by Rolf Stein in 1985 [Wikipedia 2018]. Finally, a 15th was found in 2015 by Casey Mann, Jennifer McLoud, and David Von Derau, using computer search techniques [Mann 2015]. A 2017 proof by Michael Rao, which is not yet fully peer reviewed, indicates that there are no more [Rao 2017]. Eight of the fifteen are edge-to-edge. All 15 are shown in Figure 2.17; for many of these, the pentagon shape can vary over a range of edge lengths [Scherphuis 2018].

Reinhardt also found three classes of congruent hexagons that will tile the plane, examples of which are shown in Figure 2.18. The regular tessellation of hexagons is a particular example of all three classes. It turns out there are no others, and there are no tilings by convex polygons with more than six sides.

Figure 2.17. Examples of the 15 classes of tilings by convex pentagons. The groups are outlined in black and are the smallest unit that will tile simply by translation.

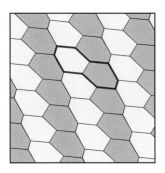 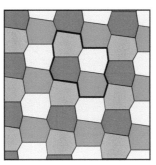 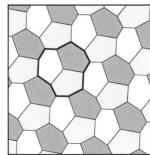

Figure 2.18. Examples of the three classes of tilings by convex hexagon tilings.

Stellation of regular polygons and star polygons

Stellation *refers to the extension of the edges of a regular polygon to form a star-like figure*. The term also applies to polyhedra in three dimensions. Some polygons allow more than one stellation, with the first stellation being the innermost, etc. In the plane, the interior of polygons can also be stellated by drawing diagonals. In this case, the first stellation is created by connecting every other corner, the second by connecting every third corner, etc. (Figure 2.19).

In geometry, more generally, the term star polygon is commonly used to refer to a self-intersecting star-shaped curve generated by connecting points evenly spaced around a circle. The notation for a star polygon generated from p points by connecting every qth point is $\{p/q\}$. Note that these figures can also be generated by the stellation of regular polygons. This sort of figure does not meet

our definition of a tile. As a result, for the purposes of this book, *the term **star polygon** will refer to star-like figures without internal lines*. The common notation for this sort of figure is $|p/q|$. Note that corners in star polygons alternatingly have convex and concave interior angles. These two types of corners can be referred to as points and dents, respectively. An n-pointed star polygon is a concave $2n$-gon.

From a graphic design standpoint, star polygons have an inherent beauty. Perhaps this is because they evoke flowers. They also have a sense of motion or energy that may be related to the way they burst or radiate outward from the center. They're widely used as religious and other symbols and are frequently seen in flags. The pentagram symbolizes the occult, the hexagram Judaism, and in conjunction with a crescent, a five-pointed star symbolizes the Ottoman Empire and Islam. The [5/2] star (Figure 2.20), known as the pentacle, figures prominently in the flags of the European Union and the United States of America.

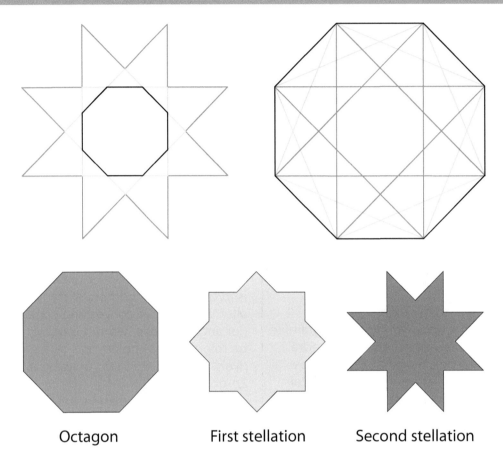

| Octagon | First stellation | Second stellation |

Figure 2.19. A regular octagon, with its first stellation and second stellation. These can be obtained either by extending the edges of the octagon or by drawing its diagonals.

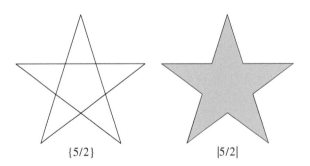

{5/2} |5/2|

Figure 2.20. Star polygons with and without internal lines. Only the latter is a tile.

Star polygon tessellations

Star polygons do not tessellate by themselves. In conjunction with other polygons, however, star polygons form a variety of beautiful tessellations. Islamic art is particularly known for complex tessellations in which star polygons feature prominently [El-Said 1976], examples of which are shown in Figure 2.21. These typically include irregular polygons as well.

There are only four edge-to-edge tessellations of star and regular polygons where every corner of a tile is also a vertex of the tessellation and for which every vertex is of the same type (Figure 2.22). There are 19 other tessellations for which every vertex is of the same type, but not every corner of a tile is a vertex of the tessellation (Figure 2.23). In addition, there are many beautiful tessellations of star and regular polygons with more than one type of vertex [Myers 2009]. Three examples are shown in Figure 2.24.

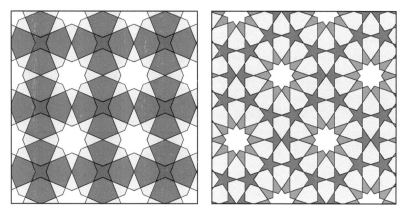

Figure 2.21. Two examples of traditional Islamic tilings incorporating star polygons.

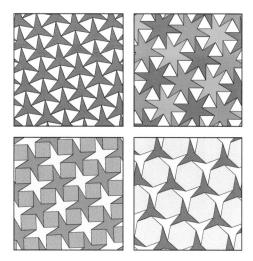

Figure 2.22. The four edge-to-edge tessellations of star polygons with regular polygons for which every vertex is of the same type and every corner of a tile is also a vertex of the tessellation.

Figure 2.23A The 19 tessellations of star polygons with regular polygons for which every vertex is of the same type, but not every corner of a tile is a vertex of the tessellation.

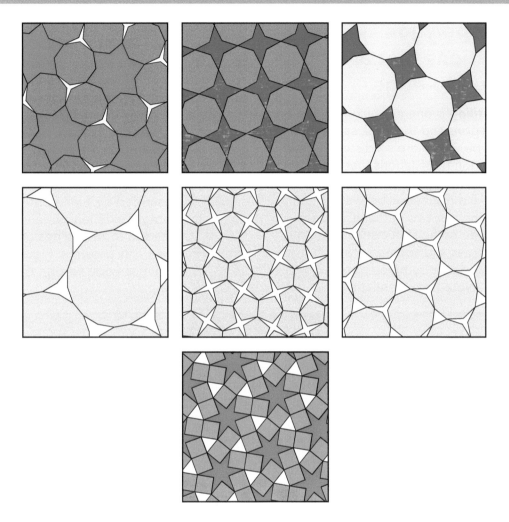

Figure 2.23B Continued

Figure 2.24. Some edge-to-edge tessellations of star and regular polygons with two types of vertices.

Regular polygon tessellations that are not edge-to-edge

A **uniform tiling** is one in which the tiles are regular polygons and the vertices are all of the same type. There are six distinct types of regular polygon tessellations that are uniform but not edge-to-edge (Figure 2.25). Some authors make a distinction between two versions of the first two of these, based on the magnitude of the shift between adjacent rows of triangles and squares, respectively. This can change the symmetry group of the tessellation. Note that the relative sizes of the different polygons can vary over some range of values for the third through sixth types. *The tessellation with two sizes of squares is often seen in real-world tiling and is sometimes referred to as the* **Pythagorean tiling**.

There are many tilings by regular polygons of multiple sizes that are not edge-to-edge. A few examples are shown in Figure 2.26. An arbitrarily large patch of square tiles of different sizes can be constructed using a method described by Alan Ponting [2014]. Squares of integer side length ranging from 1 to n^2 are used to form patches of the sort shown for $n = 19$, as shown in Figure 2.27, using Mathematica code written by Ed Pegg, Jr. [Pegg 2016].

Figure 2.25. The six distinct types of uniform tessellations of regular polygons that are not edge-to-edge.

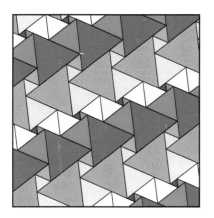
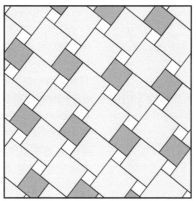
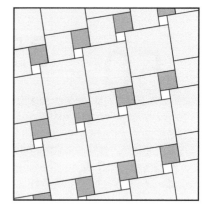

Figure 2.26. Tilings by similar regular polygons of multiple sizes that are not edge-to-edge.

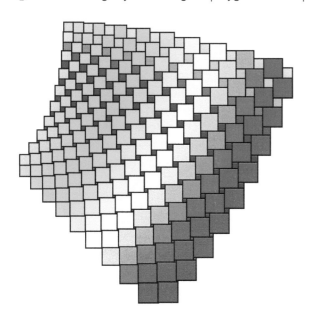

tessellations based on squared squares can be constructed by any method that tiles the plane with squares, though. The smallest squared square contains 13 smaller squares, ranging in edge length from 1 to 12 (Figure 2.28). This sort of squared square is deemed imperfect due to the use of the same size square more than once.

Figure 2.27. A patch of 361 squares with integer edge lengths from 1 to 361.

Squared squares

The mathematical recreation of tiling a square with smaller squares with edge lengths that are all integers is known as **squaring the square** *[Anderson 2016]. Since it doesn't cover the entire mathematical plane, a squared square isn't a tessellation. Infinite*

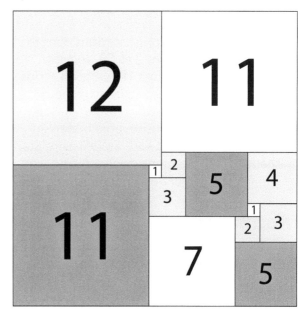

Figure 2.28. The smallest imperfect squared square. The smallest square has area 1, and all other squares have edge lengths that are integer multiples of 1.

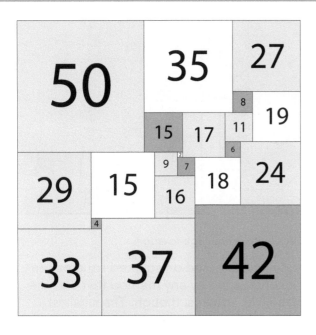

Figure 2.29. The smallest perfect squared square.

A perfect squared square doesn't use any size square more than once. The fewest number of squares required to accomplish such a tiling of a square is 21, as shown by Duijvestijn in 1978 [Anderson 2016]. The edges' lengths of the squares vary from 2 to 50, and the squared square measures 112 units on a side (Figure 2.29). Work has also been done on squared rectangles, in which rectangles are filled with squares. Considerable material about problems of this sort, as well as related mathematical recreations involved tiling, can be found at mathpuzzle.com [Pegg 2019].

Modifying tessellations to create new tessellations

There are many ways to create a new tessellation from the starting tessellation of polygons. One that was discussed above is taking the dual of the tessellation. If the starting tessellation is composed of regular polygons, creating its dual is straightforward, as each new vertex is located in the center of a polygon whose center is readily locatable. For tessellations containing other types of tiles, it can be less straightforward. A process that is more simply applied to a wider variety of tessellations is **rectification**, *in which the midpoints of each edge are connected to their nearest neighbors, followed by erasure of the original edges. If a rectified tiling is rectified a second time, the tiling is said to be* **cantellated** (Figure 2.30). Note that cantellation will convert each starting polygon in the initial tiling to a smaller similar polygon centered on the same point. There will be additional tiles introduced between these scaled-down polygons. Parallelograms are "self cantellating", meaning the cantellated and the original tessellation are the same. Because tessellations are infinite in extent, the scaling down of the polygons doesn't matter. Note that rhombuses, rectangles, and squares share this property because they are all special cases of parallelograms.

In **truncation**, *each vertex is removed and replaced by a new tile.* The number of edges of existing convex-polygon tiles is doubled. In a uniform truncation, all the new edges have the same length. While truncation works well for regular tessellations, as seen in Figure 2.31, it generally gets messier for more complicated tessellations, with edges of different lengths and irregular polygons. Rectification, cantellation, and truncation are well-known operations on polyhedra. If a polyhedron is considered to be a tiling, in which the angles meeting at each vertex sum up to less than 360°, then the usage of these terms is essentially the same for tessellations and polyhedra.

Truncation of a tiling or a polyhedron is normally assumed to mean corner truncation. However, *truncation of edges can also be performed, which is known as* **chamfering**.

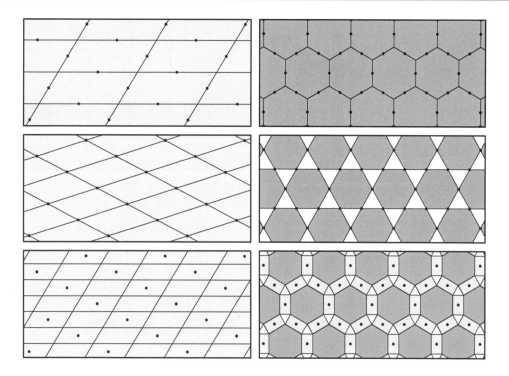

Figure 2.30. Rectification and cantellation of two tessellations. The top row shows the starting tessellations, with dots indicating midpoints of edges. The second row shows the rectified tessellations created by connecting these dots. The third row shows the cantellated tessellations created by a second rectification, along with the original set of midpoints.

Chamfering of the regular hexagon tiling is shown in stages in Figure 2.32. Chamfering to the point where all edges are of the same length, as in this case, leads to the regular hexagon tiling again, while chamfering to the point where the original tiles disappear results in a tessellation of 60°–120° rhombuses.

Compound polyhedra are superimposed geometric solids. The same concept can be applied to tessellations. Analogous to the

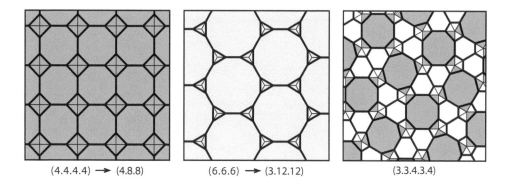

$(4.4.4.4) \rightarrow (4.8.8)$ $\qquad$ $(6.6.6) \rightarrow (3.12.12)$ $\qquad$ $(3.3.4.3.4)$

Figure 2.31. Truncation of two of the regular tessellations, creating two semi-regular tessellations, and truncation of a semi-regular tessellation.

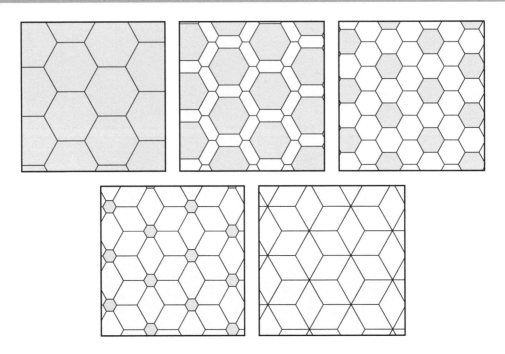

Figure 2.32. Chamfering of the regular hexagon tessellation to various degrees.

three-dimensional case, a dual compound tessellation would be a tessellation formed by overlaying a tessellation and its dual. Recall that these duals are known as the Laves tessellations and are shown earlier in this chapter. Three edge-to-edge kite tilings of this sort, which are three colored, are shown in Figure 2.33.

Another way to modify tessellations is to remove or add edges. Removing edges combines tiles to make larger ones. Adding edges creates additional tiles. There are many ways to do this for a given tessellation, and selective addition and removal of edges can be combined. Removing edges from regular-polygon tessellations is one way to generate

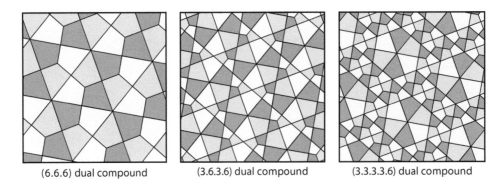

(6.6.6) dual compound   (3.6.3.6) dual compound   (3.3.3.3.6) dual compound

Figure 2.33. Tessellations formed by superimposing Archimedean tessellations with their duals.

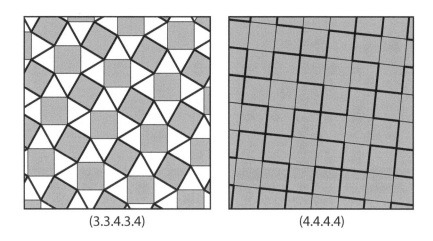

(3.3.4.3.4) (4.4.4.4)

Figure 2.34. Removal of edges from Archimedean tessellations to generate a tessellation of four-pointed stars and squares and a tessellation of X pentominoes.

star polygon tessellations. Removing edges from the regular tessellations is a way to generate polyomino, polyhex, and polyiamond tessellations (Figure 2.34).

Adding edges can create tessellations with more and different types of polygons. For example, in **triangulation**, *polygons are converted into triangles by adding edges.* There are typically many ways of doing this, as illustrated in Figure 2.35. In addition, some polygons can be left with more than three sides, creating a tessellation that is partially triangulated.

Stellation of a portion or all of the convex polygons in a tessellation creates star polygon tessellations that are in many cases quite beautiful, as the examples of Figure 2.36

illustrate. Note that the lower starting tessellation comprises irregular pentagons, making the star polygons irregular as well.

Another technique is to simply move vertices to vary the shape of polygons in a tessellation. Figure 2.37 illustrates this approach applied to two semi-regular tessellations. From a graphic design standpoint, these less regular tessellations are more whimsical and have more movement. Mirror-symmetric designs tend to look static, so removing mirror symmetry can breathe life into a design. Finally, a tessellation can be modified by applying an overall distortion. Examples of stretching and skewing Escheresque tessellations are given in Chapter 13.

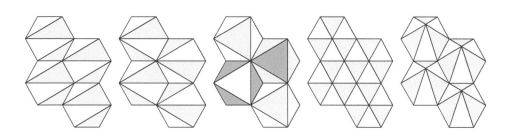

Figure 2.35. Some different ways to triangulate a patch of four regular hexagons.

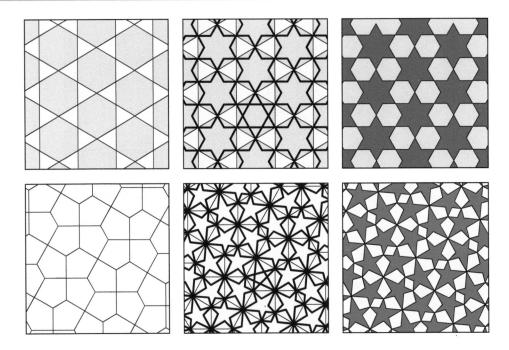

Figure 2.36. Stellation of polygons with more than four sides to create new tessellations.

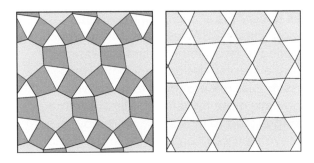

Figure 2.37. Two tessellations created by moving vertices to distort regular polygons in Archimedean tessellations.

Circle packings and tessellations

There are numerous theorems and problems related to **circle packing** – *arranging circles that touch tangentially to fill the infinite plane or a figure such as a square* [Specht 2018]. This is usually not thought of as tiling. However, the spaces between circles can be considered tiles, and many beautiful and interesting tessellations can be constructed based on circle packings. The most basic ways to pack circles in a plane are square and hexagonal packing (Figure 2.38). Note that **hexagonal packing** *is the most efficient way possible to pack circles of the same size*, with a packing density (fraction of the area covered by circles) of $\pi\sqrt{3}/6 \approx 0.9069$.

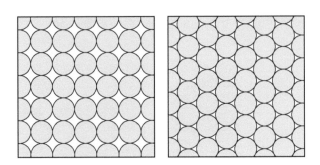

Figure 2.38. Tessellations based on square and hexagonal packings of circles.

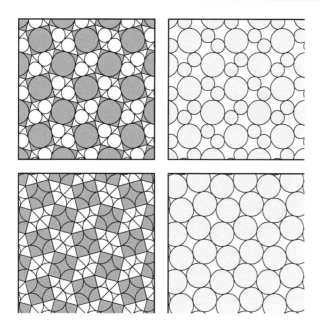

Figure 2.39. Two methods for creating tessellations with circular tiles from a semi-regular tessellation.

A variety of tessellations with circular tiles can be created from the Archimedean tessellations. Two ways of doing this are illustrated in Figure 2.39. In the first, circles are inscribed in each regular polygon. Since the inscribed circles touch each edge of every polygon at its midpoint, a tessellation of tangentially touching circles results. Each type of polygon generates a differently sized circle, and the tiles formed by the spaces in between the circles are all congruent as a result of the fact that every vertex is the same in an Archimedean tessellation. A second approach is to center a circle, with a radius equal to half the edge length, at each

vertex of the Archimedean tessellation. The circles will touch at the midpoints of each edge. While all the circles are of the same size, there will be a different tile type formed by the spaces between the circles for every type of regular polygon. Note that any tessellation of tangentially touching circles and tiles formed by the spaces between them will be two-colorable, as every vertex will have valence 4.

Complex and beautiful tessellations of circles can be generated using Descartes' circle theorem. The curvature of a circle is simply the inverse of its radius, i.e., $c = 1/r$. Given any three mutually tangent circles, the theorem provides two solutions for a fourth. If the first three have curvatures c_1, c_2, and c_3, the two solutions for c_4 are given by $c_4 = c_1 + c_2 + c_3 \pm 2(c_1c_2 + c_2c_3 + c_3c_1)^{1/2}$. As an example, if −6, 11, and 14 are c_1, c_2, and c_3, the two solutions for c_4 are 15 and 23. The negative curvature means the circle surrounds the other three mutually tangent circles. An Apollonian packing is a circle filled with mutually tangent circles. Note that the sum of the two solutions is simply $2(c_1 + c_2 + c_3)$. As a result, every circle in an Apollonian packing will have an integer curvature if the largest four have an integer curvature. The values −6, 11, and 14 were chosen so that the two solutions for c_4 would be integers. Try calculating the curvatures of the next five largest circles shown on the left of Figure 2.40. The same packing is shown on the right with a large number of circles filled in. The more circles are added, the larger the fraction of the large circle filled by circles, with the fraction going to 1 in the limit.

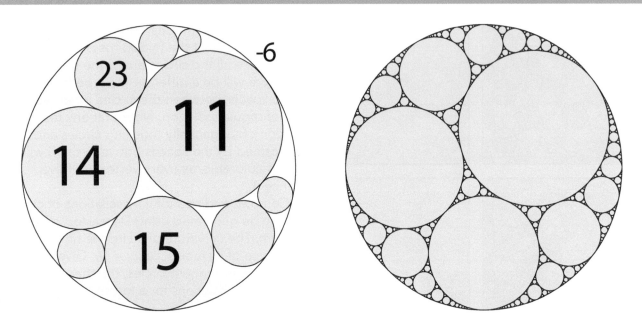

Figure 2.40. A circle can be tiled with an infinite number of smaller circles, creating a geometric object known as an Apollonian packing.

Activity 2.1. Basic properties of tiles

Materials: Copies of Worksheet 2.1.

Objective: Learn to identify tiles and their edges and corners.

Vocabulary: Tile, polygonal tile, edge, corner.

Activity Sequence

1. Write the vocabulary terms on the board and discuss the meaning of each one.
2. Pass out copies of the worksheet.
3. For each shape, have the students circle those that are tiles and X through those that aren't.
4. For the polygonal tiles, have the students put a C by each corner and an E by each edge.

Discussion Questions

1. Is the shape a/b/c/d/e/f/g/h a tile or not?
2. Why or why not?

Worksheet 2.1. Basic properties of tiles

*For each shape, circle those that are tiles and X through those that aren't.
For those that are, put a C by each corner and an E by each edge.*

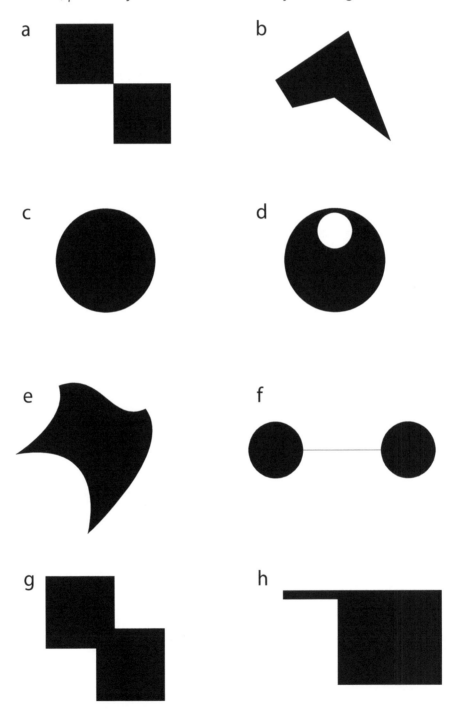

a

b

c

d

e

f

g

h

Activity 2.2. Edge-to-edge tessellations

Materials: Copies of Worksheet 2.2.

Objective: Learn to identify edge-to-edge tessellations.

Vocabulary: Edge-to-edge tessellation, vertex.

Activity Sequence

1. Write the vocabulary terms on the board and discuss the meaning of each one.
2. Pass out copies of the worksheet.
3. Have the students write an E in the margin beside each tessellation that is edge-to-edge.
4. Have the students circle one of each distinct vertex in each tessellation and number the distinct types as V1, V2, etc.

Discussion Questions

1. Is tessellation a/b/c/d/e/f/g/h edge-to-edge or not?
2. Why or why not?
3. How many distinct types of vertices does tessellation a/b/c/d/e/f/g/h possess? Describe them.

Worksheet 2.2. Edge-to-edge tessellations

Eight tessellations of pentagons or hexagons are shown here.

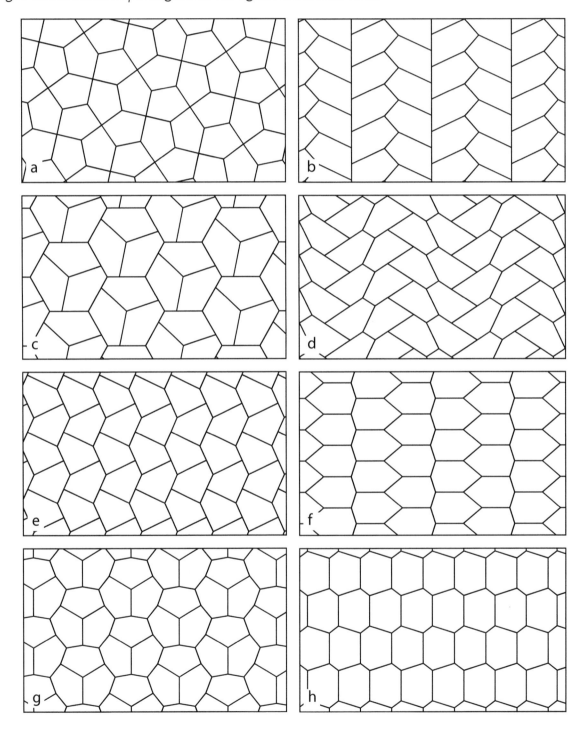

Activity 2.3. Classifying tessellations by their vertices

Materials: Copies of Worksheet 2.3.
Objective: Learn to label vertex types and use those labels to classify tessellations.
Vocabulary: Regular polygon, regular tessellation, semi-regular tessellation.
Activity Sequence
1. Write the vocabulary terms on the board and discuss the meaning of each one.
2. Pass out copies of the worksheet.
3. Have the students label each tessellation with a label that describes the vertex type.
4. Have them try to draw the tessellation described by the label at the bottom right.

Discussion Questions
1. What is the label for tessellation b/c/d/e/f?
2. Why?
   Have a student draw his or her tessellation described by (3.6.3.6) on the board.
3. Is this tessellation consistent with the label (3.6.3.6)? Why or why not?
4. Did anyone draw a different tessellation? If so, would you like to draw it on the board? Is it consistent with the label (3.6.3.6)?
5. How many different tessellations are there that are consistent with this label?

Worksheet 2.3. Classifying tessellations by their vertices

For each of these regular or semi-regular tessellations, write a label that describes the vertex type, as shown in the example. At the bottom right, try to draw the tessellation described by the label.

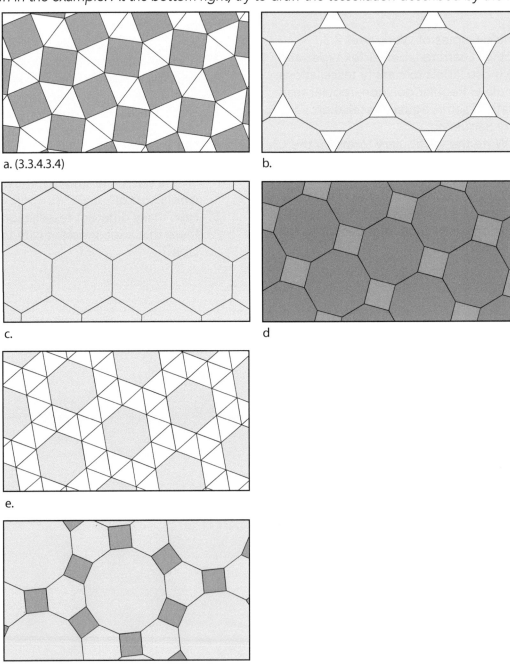

a. (3.3.4.3.4)

b.

c.

d

e.

f.

(3.6.3.6)

Chapter 3

Symmetry and Transformations in Tessellations

Symmetry and transformations are at the heart of tessellations. Fortunately, it doesn't require a lot of mathematics background to gain a basic understanding of them. A transformation is an act of moving or changing one thing into another. Symmetry is the property of invariance under a transformation, i.e., not being altered by the transformation. In a repeating pattern, there are only certain types of symmetry that are possible in one, two, and three dimensions. Before exploring the mathematical classification of these, symmetry and transformations will be described using real-world objects as examples.

Symmetry in objects

Strictly speaking, an isolated object can only possess two basic types of symmetry – mirror (reflection) and rotational. However, an object can also possess translational or glide reflection symmetry in a local sense.

*An object possesses **mirror symmetry** if it can be reflected in some line (in two dimensions; plane in three dimensions) passing through its center and remain unchanged.* An object can possess multiple lines of mirror symmetry. *A particular type of mirror symmetry is **bilateral symmetry**, which means that an object has a single line or plane of mirror*

Figure 3.1. Three examples of real-world objects that possess bilateral symmetry about a center line (actually a plane in three dimensions).

symmetry (see examples in Figure 3.1). Most animals have approximate bilateral symmetry. Real-world objects are never mathematically exact, so it should be understood when referring to symmetry in real objects that such symmetry is approximate. This is especially true of objects occurring in nature.

An object possesses **rotational symmetry** *if it can be rotated by some fraction of a full revolution about its center and remain unchanged.* If that fraction is 1/*n* of a full revolution, that object is said to possess *n*-fold rotational symmetry. An object can possess both reflection and rotational symmetry, as in the sea star of Figure 3.2. In two dimensions, the rotation center is a point, while in three dimensions it's a line. Additional examples of symmetrical objects are shown in Figure 3.3.

An object possesses **translational symmetry** *if it can be translated (moved by some amount in some direction) and remain unchanged.* Strictly speaking, a physical object cannot possess translational symmetry, because physical objects are finite in extent. A translation would therefore move part of the object of its original footprint. However, over a limited region, an object can possess translational symmetry. This sort of local translational symmetry is not uncommon, as shown in Figure 3.4.

An object possesses **glide reflection symmetry** *if it can be translated along some line (glide) and then be reflected about that line and remain unchanged.* Strictly speaking, a physical object cannot possess glide reflection symmetry, since a translation would move part of the object of its original footprint. However,

Figure 3.2. This sea star has five-fold rotational symmetry, as shown, as well as mirror symmetry about five different lines, one of which is indicated by dashes.

Figure 3.3. This flower has five-fold rotational symmetry, but no lines of mirror symmetry. This wheel possesses mirror symmetry about the dashed line but does not possess rotational symmetry. The nine spokes and the five bolts alone possess rotational symmetry, but the wheel as a whole does not.

Figure 3.4. The window and tiles exhibit local translational symmetry in two dimensions, while the building frieze exhibits local translational symmetry in one dimension (horizontally).

Figure 3.5. Two examples of real-world objects possessing local glide reflection symmetry.

within a limited region, an object can possess glide reflection symmetry (Figure 3.5).

Transformations

There are four types of transformations a two-dimensional figure can make within the plane that could leave it unchanged. These are translation, rotation, reflection, and glide reflection. Reflection can be considered a special case of glide reflection in which the glide distance is zero.

The simplest transformation is translation, sometimes called a slide or a glide, which is movement from one location to another with no rotation or reflection. If an object such as a polygon translates, both distance and direction are associated with that movement (Figure 3.6). *An arrow indicating both distance and direction is called a* **vector**.

Two edges of a tile can be related to each other by a translation. In the templates later in this book, this transformation is indicated by a light-blue arrow, such as in Figure 3.7. This

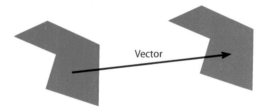

Figure 3.6. The translation of an object is described by a vector, an arrow indicating both direction and distance.

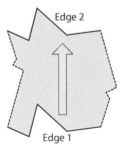

Figure 3.7. The two solid lines in this tile are related by a translation that transforms Edge 1 into Edge 2.

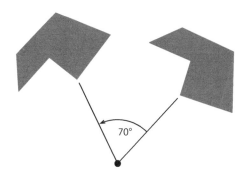

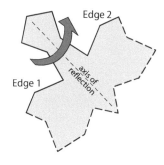

Figure 3.8. The rotation of an object is specified by an angle of rotation and a point about which the rotation takes place.

Figure 3.10. The two solid lines in this tile are related by a reflection about the dashed line bisecting the tile that transforms Edge 1 into Edge 2. Imagine the arrow forming a sort of bridge above the axis of reflection.

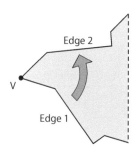

Figure 3.9. The two solid lines in this tile are related by a rotation about the point V that transforms Edge 1 into Edge 2.

arrow simply indicates the type of transformation and its direction, not the precise distance.

When an object moves within the plane by rotating without translating or reflecting, that movement can be fully specified by a center of rotation and an angle of rotation between

0° and 360° (Figure 3.8). In the templates later in this book, two edges of a tile are often related to each other by rotation. This is indicated with a curved arrow of the sort shown in Figure 3.9.

A reflection is specified simply by a line of mirror symmetry. In the templates later in this book, if two edges of a tile are related by reflection, the axis of reflection is indicated by a dashed line, with an arching arrow illustrating the transformation between the two edges, as in Figure 3.10.

A glide reflection involves two motions, a translation (or glide) and a reflection. In the templates later in this book, an arrow twisted to 180° is used to indicate a glide reflection transformation between the edges at either end of the arrow, as in Figure 3.11.

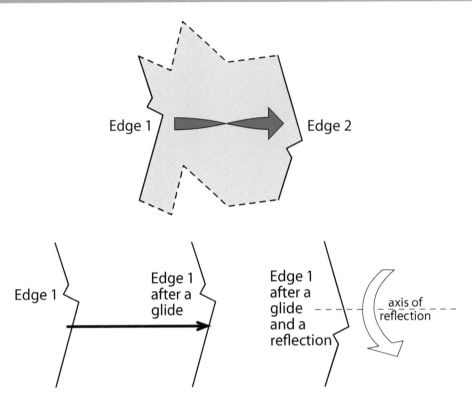

Figure 3.11. Edge 1 is transformed into Edge 2 by the combination of a glide and a reflection, which can be indicated schematically by a twisted arrow.

Symmetry in tessellations

*A tessellation is said to possess **translational symmetry** if the infinite tessellation can be translated and remain unchanged (perfectly overlie itself). A tessellation that possesses translational symmetry is said to be **periodic**.* The translation vector for an infinite tessellation is not unique, as illustrated in Figure 3.12.

*A tessellation is said to possess n-fold **rotational symmetry** about a point (where n is an integer) if the entire tessellation can be rotated by 1/n of a full revolution about that point and remain unchanged.* The smallest rotation amount should be specified; i.e., if a tessellation possesses six-fold rotational symmetry, it also possesses three-fold and two-fold, but it is referred to as six-fold. A polygon can be used both to indicate a point

of rotational symmetry and to specify the amount of rotation. For example, a rectangle is used to indicate a point of two-fold rotational symmetry, an equilateral triangle of three-fold, a square of four-fold, a hexagon of six-fold, and so on. The tessellation of Figure 3.13 demonstrates the marking of three- and four-fold rotation points.

*A tessellation is said to possess **glide reflection symmetry** if the entire tessellation can be translated along some line and then reflected about that line and remain unchanged.* Mirror symmetry is a special case of glide reflection symmetry, in which the glide distance is zero. In this book, mirror symmetry is usually indicated with a line of short dashes, while glide reflection symmetry is indicated by a line of long dashes (Figure 3.14). In a periodic tessellation, there are an

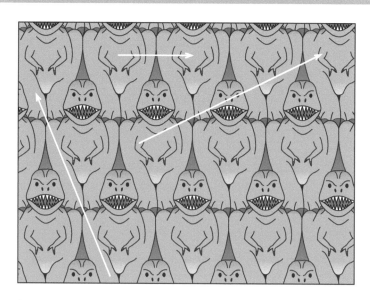

Figure 3.12. Any of the white arrows can translate this tessellation of T-Rexes without changing it.

Figure 3.13. This horned lizard and Gila monster tessellation contains one distinct point of six-fold rotational symmetry and two distinct points of three-fold rotational symmetry, indicated by a hexagon and equilateral triangles, respectively. If the interior details are ignored, there are also points of two-fold rotational symmetry at the center of each Gila monster.

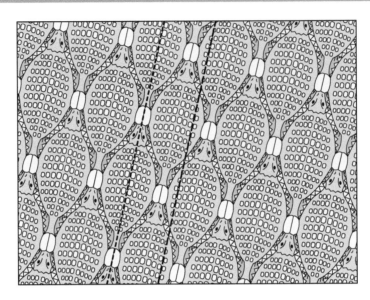

Figure 3.14. This tessellation of ankylosaurus possesses one distinct line of mirror symmetry and one distinct line of glide reflection symmetry, indicated by short dashes and long dashes, respectively.

infinite number of equally spaced lines of a given type, unless the tessellation is only periodic in the direction parallel to the line.

Frieze groups

The collection of all symmetries possessed by a given tessellation or other pattern specifies its **symmetry group**. Before considering the symmetry groups of patterns that repeat in two directions, the symmetry groups of *patterns that repeat in only one direction, known as* **friezes**, will be described. There are many common uses of friezes, including architectural elements, borders for rugs and carpet, borders for calligraphy or certificates, baskets and pots, dishes, and clothing elements, as shown in Figure 3.15. Many friezes are based on abstract geometric motifs, but certain representational motifs seem to lend themselves well to use in friezes. Botanical themes are particularly popular, especially leaves, vines, and flowers. Knot work has also been used extensively in friezes, and botanical elements like vines can be interwoven in a knot-like fashion.

Frieze groups provide a classification of friezes based on the symmetries they possess. It has been proven mathematically that there are only seven possibilities for the symmetries of a pattern that repeats in one direction, thus, the seven frieze groups. For the purpose of discussing the symmetries of a frieze pattern, the pattern is assumed to extend indefinitely in the repeat direction.

Frieze groups are commonly described by a four-character international symbol. The first character is *p* (for primitive cell) for all seven groups, and the second character is either *m*, indicating the presence of lines of mirror (reflection) symmetry perpendicular to the repeat direction, or 1, indicating the lack of such reflections. The third character may be *m*, indicating a line of reflection symmetry along the center line in the repeat direction, 1, indicating the lack of such reflection, or *g*, indicating a line of glide reflection symmetry (but not simple reflection) along the center line. The final character may be either 1, indicating no rotational symmetry, or 2, indicating points of two-fold rotational symmetry. Other notations are used as well, so the labeling of frieze patterns is not always the same from one source to another. In the next paragraph, the orbifold signature for each group is provided in parentheses following the international symbol. The orbifold system is described briefly later in this chapter.

Simple graphical examples of the seven frieze groups are shown in Figure 3.16. The simplest group is $p111$ ($\infty\infty$), with no

symmetries other than translation in one direction. Patterns in the frieze group $pm11$ ($*\infty\infty$) additionally contain lines of reflection perpendicular to the repeat direction. In contrast, $p1m1$ ($\infty*$) friezes contain a reflection about the center line along the repeat direction. Patterns in the frieze group $p1g1$ ($\infty\times$) contain a line of glide reflection, but no simple reflections or rotations. Friezes in group $p112$ (22∞) possess two-fold rotational symmetry, but no reflections or glide reflections. Both the $pmg2$ and $pmm2$ groups have

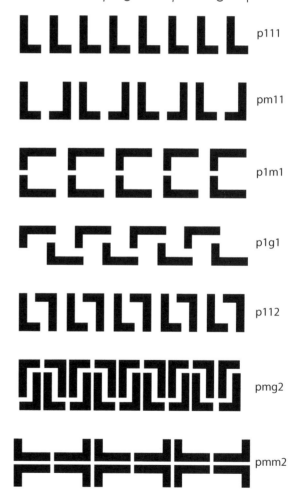

Figure 3.15. Examples of real-world friezes: top to bottom, two architectural friezes in Riga, Latvia; a design on the spine of a book; and a detail from a Bank of Scotland twenty pound note.

Figure 3.16. The seven frieze groups, which describe the possible symmetries of patterns that repeat in one direction.

reflections perpendicular to the repeat direction and two-fold rotational symmetry. The *pmg2* (2*∞) group also has glide reflection along the repeat direction, while the *pmm2* (*22∞) group has a simple reflection about the center line along the repeat direction. Which groups do the examples of Figure 3.15 belong to?

Wallpaper groups

The collection of all symmetries possessed by a pattern that repeats in two directions specifies its **plane symmetry group** *or* **wallpaper group**. There are 17 distinct wallpaper groups, and each group is given a unique descriptive label. Examples of symmetry group international symbols are *p3*, *p4g*, and *cmm*. The leading *p* or *c* stand for primitive cell or face-centered cell. If there is a number in the second position, it indicates the highest order of rotational symmetry. The subsequent *m*, *g*, or 1 (there can be more than one *m* or *g* for different axes) indicate mirror, glide reflection, or neither, as for the frieze groups. My own examples of Escheresque tessellations for all 17 are shown in Figure 3.17. The coloring should be ignored as far as determining the group is concerned, as differing colors are used to make the motifs stand out better. The group name is given in the short notation in the figure. The orbifold notation is given below in regular parentheses.

In Figure 3.17, *the* **translation unit cell**, *the smallest portion that will generate the full pattern by translation only, is indicated by a bold outline. The* **primitive unit cell**, *the smallest portion that will generate the full pattern by any combination of translations, rotations, and glide reflections, is colored in white. Simple reflections are indicated by dotted lines, while glide reflections are* indicated by dashed lines. Points of two-fold rotational symmetry are indicated by rectangles. If there are two or more distinct points of two-fold symmetry, this is indicated by the rectangles having different orientations. Points of three-fold rotational symmetry are indicated by triangles, and triangles for distinct points have different orientations. Points of four-fold rotational symmetry are indicated by squares, and squares for distinct points also have different orientations. Points of six-fold rotational symmetry are indicated by hexagons. Figure 3.18 shows a chart that can aid in determining a pattern's symmetry group [Wikipedia 2019].

p1 (O)

This is the simplest wallpaper group, containing only translations. The primitive unit cell and translation unit cell are the same.

p2 [p211] (2222)

In addition to translations, patterns in this group have four distinct points of two-fold rotational symmetry. There are two primitive unit cells in the translation unit cell.

pm [p1m1] (**)

Patterns in this group have two distinct, simple reflections that are parallel and no rotations. There are two primitive unit cells per translation unit cell.

pg [p1g1] (XX)

Patterns in this group have two distinct glide reflections that are all parallel and no simple reflections. There are two primitive unit cells per translation unit cell.

Figure 3.17. The 17 plane symmetry groups illustrated with Escheresque tessellations. The coloring is not taken into account in determining the symmetry groups.

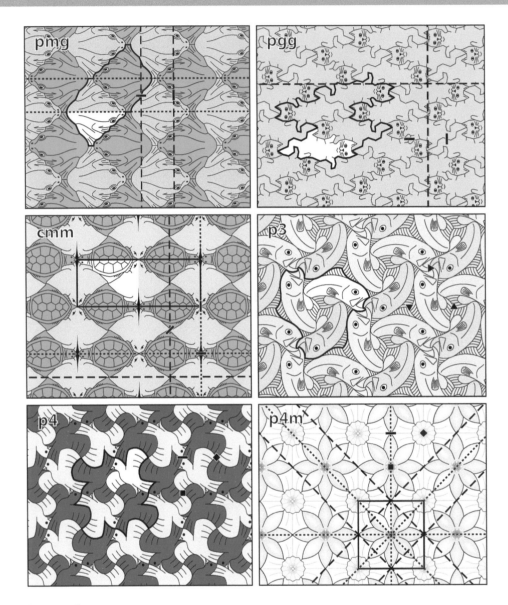

Figure 3.17. Continued.

cm [c1m1] (*X)

Patterns in this group have one distinct, simple reflection that is parallel to one distinct glide reflection. The glide reflection lines fall midway between the simple reflection lines. There are two primitive unit cells per translation unit cell.

pmm [p2mm] (*2222)

Patterns in this group have simple reflections in perpendicular directions and four distinct two-fold rotation centers at the intersections of the reflection lines. There are four primitive unit cells per translation unit cell.

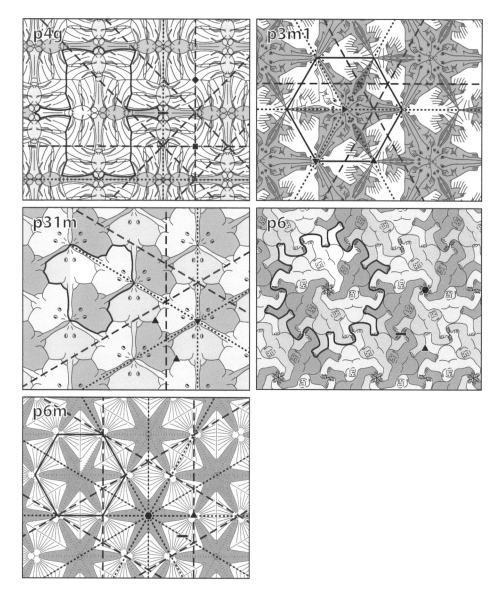

Figure 3.17. Continued.

pmg [p2mg] (22*)

Patterns in this group have two distinct centers of two-fold rotation, mirror lines in one direction, and glide reflection lines perpendicular to those. The rotation centers lie on the glide reflection axes. There are four primitive unit cells per translation unit cell.

pgg [p2gg] (22X)

Patterns in this group have two distinct centers of two-fold rotation, no simple reflections, and two glide reflection lines perpendicular to one another. The rotation centers do not lie on the glide reflection axes. There are four primitive unit cells per translation unit cell.

Smallest rotation	Does the pattern have simple reflections?		
	Yes		No
360°/6	p6m		p6
360°/4	Does it have reflections at 45°?		p4
	Yes: p4m	No: p4g	
360°/3	Does it have rotation centers off reflections?		p3
	Yes: p31m	No: p3m1	
360°/2	Does it have perpendicular reflections?		Does it have glide reflections?
	Yes	No	Yes: pgg No: p2
	Rotation centers off reflections?	pmg	
	Yes: cmm No: pmm		
none	Does it have a glide axis off mirrors?		Does it have glide reflections?
	Yes: cm	No: pm	Yes: pg No: p1

Figure 3.18. Chart for determining to which of the 17 plane symmetry groups a pattern belongs, by starting with the top question and work your way down.

cmm [c2mm] (2*22)

Patterns in this group have three distinct centers of two-fold rotation, mirror lines in two perpendicular directions, and glide reflection lines in two perpendicular directions. Two of the rotation centers lie at the intersections of the simple reflection axes, and one at the intersections of the glide reflection axes. There are four primitive unit cells per translation unit cell.

p3 (333)

Patterns in this group have three distinct centers of three-fold rotation, no simple reflections, and no glide reflections. There are three primitive unit cells per translation unit cell.

p4 (442)

Patterns in this group have one center of two-fold rotation, two distinct centers of four-fold rotation, no simple reflections, and no glide reflections. There are four primitive unit cells per translation unit cell.

p4m [p4mm] (*442)

Patterns in this group have one center of two-fold rotation, two distinct centers of four-fold rotation, simple reflections in four directions, and two glide reflections. The rotation centers all lie at the intersection of the reflection axes. There are eight primitive unit cells per translation unit cell.

p4g [p4mg] (4*2)

Patterns in this group have one center of two-fold rotation, two centers of four-fold rotation, simple reflections in two perpendicular directions, and glide reflections in four directions. The two-fold rotation centers lie at the intersection of the simple reflection axes. There are eight primitive unit cells per translation unit cell.

p3m1 (*333)

Patterns in this group have three distinct centers of three-fold rotation, simple reflections in three directions, and glide reflections along the three lines lying halfway between the simple reflections. The rotation centers lie at the intersection of the simple reflection axes. There are six primitive unit cells per translation unit cell.

p31m (3*3)

Patterns in this group have three distinct centers of three-fold rotation, two of which are mirrors of each other (marked with the same orientation of triangle in Figure 3.17), simple reflections in three directions, and glide reflections along the three lines lying halfway between the simple reflections. At least one of the rotation centers does not lie on a reflection axis. There are six primitive unit cells per translation unit cell.

p6 (632)

Patterns in this group have one center each of six-fold rotation, three-fold rotation, and two-fold rotation and no simple reflections and glide reflections. There are six primitive unit cells per translation unit cell.

p6m [p6mm] (*632)

Patterns in this group have one center each of six-fold rotation, three-fold rotation, and two-fold rotation and simple reflections in six directions and glide reflections along the six lines that lie midway between the simple reflection axes. There are 12 primitive unit cells per translation unit cell.

Heesch types and orbifold notation

There are several other systems for classifying symmetries, two of which will be described here.

Orbifold notation is a more recent symmetry classification system that is popular with mathematicians due to some inherent advantages over earlier notations. This system won't be fully explained here, but the orbifold notations (signature) for the frieze and wallpaper groups are provided in the earlier sections in parentheses. In the orbifold notation, all the symmetries of any pattern can be described by wonders, gyrations, kaleidoscopes, and miracles, which are specified in this order in the signature. The wonder-ring O represents patterns that only repeat by translation. Gyrations are specified by a number denoting the degree of rotational symmetry, including the infinity symbol ∞ for infinite rotational symmetry. Kaleidoscopic or mirror symmetry is denoted by the star * along with a numeral indicating the number of mirror lines that pass through each distinct point of the mirror symmetry. Finally, the miracle symbol × represents a glide reflection. If you're interested in learning more about this way of looking at symmetry, you might want to read *The Symmetries of Things*, by John H. Conway, Heidi Burgiel, and Chaim Goodman-Strauss [Conway et al. 2008].

Heinrich Heesch classified all of the distinct types of asymmetric tiles that will tessellate **known as Heesch types**. He found 28 distinct types in the early 1930s, but didn't publish these results until 1963. By this time, M.C. Escher had already worked out his own classification system, which included almost all of the Heesch types [Schattschneider

2004]. Escher's system has some similarities to Heesch's but is harder to understand, so it won't be described here. Heesch's system, while not widely used in general, is particularly well suited for designing Escheresque tessellations. For each of the templates provided in later chapters, the Heesch type is given when applicable. For designing Escheresque tiles, the mirror symmetry is often useful, since many living creatures have bilateral symmetry. Tiles of this sort are excluded from Heesch's classification since they are not asymmetric. In addition, Heesch's system only includes tessellations with a single prototile. In contrast, a symmetry group can be assigned to every tessellation.

Heesch's notation is not difficult to understand. It describes the tessellation in terms of the transformations relating to the edges of one of the tiles. The letter T denotes translation, G glide reflection, and C rotation. A translation must always be from one side of the tile to the opposing side, so subscripts are not needed for T. For G, there is no subscript if there is only one line that is transformed by glide reflection, and subscripts 1 and 2 to differentiate when there are two lines of a glide reflection. For C, the subscript indicates the amount of rotation. C without a subscript indicates 1/2 of a full revolution, 3 indicates 1/3, 4 indicates 1/4, and 6 indicates 1/6.

To use Heesch's system, first label each edge of a tile. The edges are always labeled in pairs except for edges with two-fold rotational symmetry. Once the edges are labeled, the tile is described by going around it in a circuit, listing the edge types. Some examples are shown in Figure 3.19, with vertices indicated by dots. (One needs to see the tessellation to see where the vertices are.) On the left of the figure, the tile has four edges, with opposing sides related by translation. Each edge is labeled T, and the Heesch type is simply TTTT. In the middle, two of the three edges are related by a glide reflection, while the third edge possesses two-fold rotational symmetry about the tick mark. The labels are then G, G, and C, and the Heesch type is CGG. In the example on the right, the six sides occur in pairs that are related by rotations of 1/3 of a full revolution. Thus each side is labeled C3, and the Heesch type is $C_3C_3C_3C_3C_3C_3$.

Coloring of tessellations and symmetry

There are many ways to color a tessellation. In most cases, adjacent tiles are given different colors so that they visually stand out more. Escher obeyed this rule in all of his

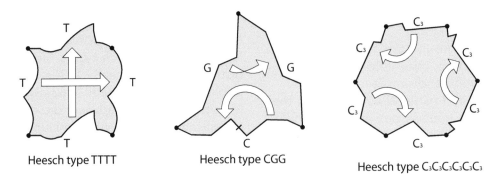

Heesch type TTTT   Heesch type CGG   Heesch type $C_3C_3C_3C_3C_3C_3$

Figure 3.19. Examples of the use of Heesch types to described tiles.

tessellations. The mathematical question here is the minimum number of colors required to accomplish this for a given tessellation. All tessellations that are Heesch types require either two or three colors. It has been proven mathematically in the four-color theorem that no more than four colors are needed for any tessellation to ensure that no two adjacent regions have the same color.

Another option is to color all tiles of a given type the same, regardless of whether or not they are adjacent. This sort of coloring was used in Figure 2.12, which shows the semi-regular tessellations.

If two colors suffice to ensure no adjacent tiles are of the same type, the tessellation is said to be two-colorable. Any tessellation for which all vertices are of even valence can be two colored. If a tessellation contains one or more vertices with odd valence, it cannot be two colored. The regular triangle and square tessellations can be two colored, but the hexagon tessellation cannot. Only two of the eight semi-regular tessellations can be two colored. These are shown in Figure 3.20, along with examples of tessellations requiring three and four colors.

Symmetry is one consideration in the choice of the coloring of tessellations. A given coloring can preserve all, some, or none of the symmetries of a tessellation. For example, if a particular glide reflection leaves an uncolored tessellation unchanged, does it still leave the colored tessellation unchanged? Visually, a choice that preserves symmetries will bring out the mathematical symmetry more.

A particular type of coloring that Escher considered was perfect coloring. In a perfectly colored tessellation, it isn't necessarily the case that a symmetry operation that leaves an uncolored tessellation unchanged will leave the colored tessellation unchanged.

Figure 3.20. Two examples each of tessellations that require two, three, and four colors to ensure no two adjacent tiles are of the same color.

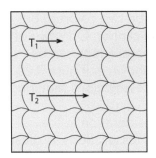
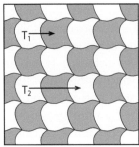

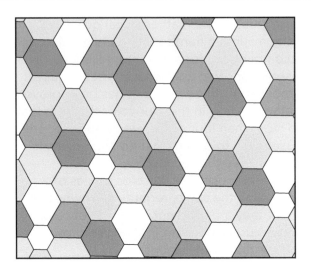

Figure 3.21. Translation T1 doesn't change the one-color tessellation, but does change the two-color one. However, every yellow tile is made red and every red tile is made yellow. Translation T2 changes neither. Since any other translation would be equivalent to one of these two, the two-coloring is perfect.

However, any changes must take the form of an unambiguous permutation (swapping out) of the colors. For example, if a tessellation has red, green, and blue tiles, a symmetry operation that moves a blue tile onto a red tile must move every blue tile onto a red tile. An example is provided in Figure 3.21. In a perfectly colored tessellation, symmetries that permute the colors, as opposed to leaving them unchanged, are called color symmetries. These are color-exchanging symmetries of the design. A mathematical question here is the minimum number of colors required to perfectly color a tessellation. This number can be

Figure 3.22. As many as seven colors may be required to preserve all translational symmetries respective of the color, with the condition that no two adjacent regions have the same color.

greater than the minimum number required to prevent adjacent tiles having the same color.

If one wishes to preserve all translational symmetries in a periodic tessellation, while avoiding adjacent tiles sharing a color, as many as seven colors could be required. This follows the seven color theorem for the number of colors needed to guarantee a map on a torus that has different colors for all adjacent regions. An example of a planar tessellation requiring seven colors is shown in Figure 3.22.

Activity 3.1. Symmetry in objects

Materials: Copies of Worksheet 3.1.
Objective: Learn to identify rotational and mirror symmetry in manmade and natural objects.
Vocabulary: Rotational symmetry, mirror symmetry, bilateral symmetry.
Specific Common Core State Standard for Mathematics addressed by the activity
    Recognize a line of symmetry for a two-dimensional figure as a line across the figure such that the figure can be folded along the line into matching parts. Identify line-symmetric figures and draw lines of symmetry. (4.G.3)

Activity Sequence
1. Write the vocabulary words on the board and discuss the meaning of each.
2. Pass out copies of the worksheet.
3. First, go over the example. Then have the students mark the lines of mirror symmetry and write down the rotational symmetry of the objects on the page.

Discussion Questions
1. How many lines of symmetry did you mark for the object a/b/c/d/e/f/g? Why?
2. Does the object a/b/c/d/e/f/g possess rotational symmetry? If so, what is the minimum rotation n that will leave the object unchanged?

Worksheet 3.1. Identifying symmetry in objects

Mark the lines of mirror symmetries and write down the rotational symmetry of the objects on this page, as shown in the example. If the object does not possess rotational symmetry, write "No rotational symmetry".

4-fold rotational symmetry

a

b

c

d

e

f

g

Activity 3.2.
Transformations

Materials: Copies of Worksheet 3.2.

Objective: Learn to identify transformations in Escheresque tiles.

Vocabulary: Translation, rotation, glide reflection.

Activity Sequence

1. Write the vocabulary words on the board and discuss the meaning of each.

2. Pass out copies of the worksheet.

3. First go over the example. Then have the students mark and label the transformations between the two edges marked "a" for each tile.

Discussion Questions

1. What sort of transformation did you mark for the tile a/b/c/d/e/f? Why?

Worksheet 3.2. Identifying transformations

For each tile below, use arrows to indicate the transformation that relates the two edges marked "a" and write the type of transformation, as shown in the examples.

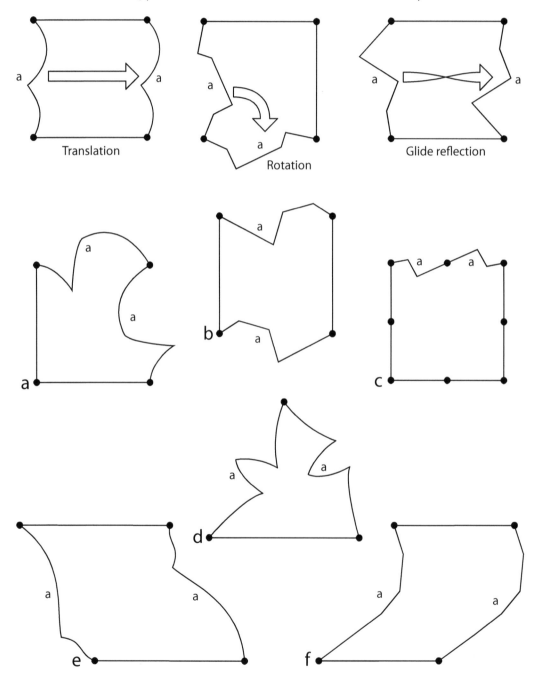

Translation

Rotation

Glide reflection

Activity 3.3. Translational symmetry in tessellations

Materials: Copies of Worksheet 3.3.

Objective: Learn to identify translational symmetry in tessellations and to mark unit cells and translation vectors.

Vocabulary: Translational symmetry, vector, unit cell.

Activity Sequence

1. Write the vocabulary words on the board and discuss the meaning of each.
2. Pass out copies of Worksheet 3.3.
3. For each tessellation, have the students draw two different translation vectors that would leave the tessellation unchanged.
4. For each tessellation, have the students outline the unit cell.

Discussion Questions

1. Share a vector you drew for tessellation a/b/c/d/e/f.
2. Does anyone think the vector shown is not a translation vector for this tessellation? If so, why?
3. How many tiles of each type are contained in the unit cell for tessellation a/b/c/d/e/f/g/h?
4. Did anyone get a different result? If so, which is correct and why?

Worksheet 3.3. Identifying symmetry in tessellations

Mark the symmetries of each tessellation as instructed by your teacher.

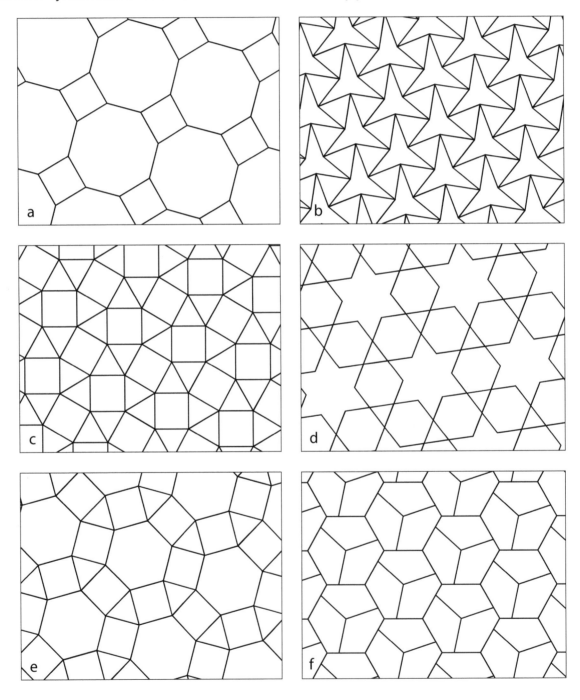

Activity 3.4. Rotational symmetry in tessellations

Materials: Copies of Worksheet 3.3.

Objective: Learn to identify rotational symmetry in tessellations and to mark rotation points with polygons indicating the amount of rotation.

Vocabulary: Rotational symmetry.

Activity Sequence

1. Write the vocabulary term on the board and discuss its meaning.
2. Pass out copies of Worksheet 3.3.
3. For each tessellation, have the students draw a polygon on each distinct point of the rotational symmetry. Have them use a rectangle for points of two-fold rotational symmetry, a triangle for three-fold, a square for four-fold, and a hexagon for six-fold.

Discussion Questions

1. Share your polygon locations and types for tessellation a/b/c/d/e/f.
2. Does anyone think any of the polygons are not the right type or at the right location? Why?
3. Did anyone find any additional distinct points of rotational symmetry for the tessellation?

Activity 3.5. Glide reflection symmetry in tessellations

Materials: Copies of the Worksheet 3.3.

Objective: Learn to identify glide reflection symmetry in tessellations and to mark glide lines and vectors.

Vocabulary: Mirror symmetry, glide reflection symmetry, vector.

Specific Common Core State Standard for Mathematics addressed by the activity
     Recognize a line of symmetry for a two-dimensional figure as a line across the figure such that the figure can be folded along the line into matching parts. Identify line-symmetric figures and draw lines of symmetry. (4.G.3)

1. Write the vocabulary terms on the board and discuss their meanings.
2. Pass out copies of Worksheet 3.3.
3. For each tessellation, have the students mark lines of mirror and glide reflection symmetry, labeling the former "M" and the latter "G" and indicating a possible glide vector for the latter.

Discussion Questions

1. Do any of the tessellations possess neither mirror nor glide reflection symmetry? Which one(s)?
2. Do any of the tessellations possess both mirror and glide reflection symmetry? Which one(s)?
3. How many distinct lines of glide reflection symmetry does tessellation a/b/c/d/e/f possess?

Chapter 4

Tessellations in Nature

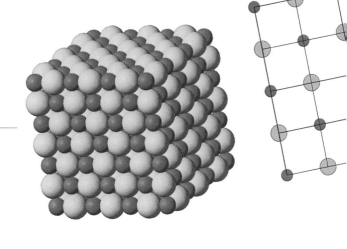

Tessellations are all around us, but most people aren't fully aware of how integral a part of our world they are. Real-world tessellations can be divided into two main categories, natural and synthetic. Tessellations in nature are often not very regular, and no real-world tessellation strictly meets the mathematical definition of a tessellation. Recall that a tessellation is a collection of shapes that fit together without overlaps or gaps to cover the infinite mathematical plane. In a real-world tessellation, the lines or gaps between the tiles will always have a finite width. The individual tiles in what is nominally a repeating tessellation will never be exactly like one another. No real-world tessellation can be infinite in extent, and no planar surface in the real world is perfectly flat.

Modeling of natural tessellations

Overlooking the fine points described above, real-word structures that approximate mathematical tessellations will be described as tessellations in this and the following chapter. The mathematical tessellation that is used to describe a physical structure is a model that approximates something in the real world. Models with a mathematical basis are widely used in sciences as a way of helping us understand the natural world. Several models of tessellations will be described in the following sections, including the use of a regular tessellation to model honeycomb, a drawing of a non-repeating tessellation of polygonal tiles with varying numbers of sides to approximate the structure of a dragonfly's wing and the construction of a particular type of tessellation to approximate the patches of color on a giraffe.

Crystals

On the atomic scale, atoms in crystals organize themselves into regular geometric lattices as a result of physical forces. *A **lattice** is a periodic arrangement of points in two or more dimensions.* Crystallography, the study of crystalline matter, played an important role in the understanding of symmetry and the notation of tiling. Real-world crystals are three dimensional (Figure 4.1a), but individual layers of atoms in crystals form lattices in two dimensions (Figure 4.1b). The lattices of crystalline materials can lead to macroscopic geometric features in crystals such as iron pyrite and fluorite (Figure 4.2).

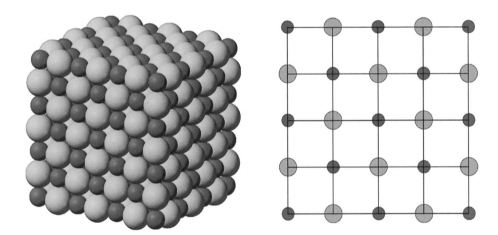

Figure 4.1. (a) The 3D lattice of atoms comprising NaCl (table salt), with Na⁺ ions shown as blue and Cl⁻ ions shown as green. (b) A single plane of atoms in a NaCl crystal, with lines connecting the nearest neighbors to form a tessellation of squares.

Figure 4.2. Geometric features of fluorite specimens are a manifestation of its atomic lattice.

Lattices

As shown in Figure 4.1, the points of a lattice in two dimensions can be connected to form tessellations. The same lattice of points can be connected in different ways to form different tessellations. The most basic two-dimensional lattices are based on the square and the equilateral triangle (Figure 4.3). The points of a triangular lattice can be connected to form a regular tessellation of hexagons, illustrating the close relationship between the regular tessellations of equilateral triangles and hexagons. *Note that **grid** is another word for a lattice in two dimensions.*

Just as lines connect points in two dimensions to define two-dimensional tessellations, surfaces connect points in three dimensions to form three-dimensional tessellations. A simple cubic lattice of points in three dimensions can be connected with planes to form cubes that tessellate three-dimensional space. The tiles in a three-dimensional tessellation are solids such as polyhedra, as will be discussed in greater detail in Chapter 20.

Cracking and crazing

Layered structures are common in nature. The breaking up of a continuous layer into individual pieces, such as the cracking of mud as it dries, can be a stress-relief mechanism (Figure 4.4). Cracking patterns can be random or have preferential directions, depending on the uniformity of the stress in the material and the homogeneity of its structure. In some cases, geometric tessellations are formed by the natural cracking of rocks (Figures 4.5 and 4.6).

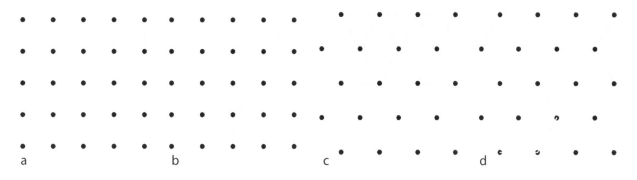

a b c d

Figure 4.3. A square lattice of points connected to form (a) a regular tessellation of squares and (b) a tessellation of parallelograms. A lattice of points defined by equilateral triangles can be connected to form (c) a regular tessellation of equilateral triangles or (d) a regular tessellation of hexagons.

Figure 4.4. Cracked mud (Hungary).

Figure 4.7. A cross section through these basalt columns would reveal a tessellation of polygons, many of them hexagons (Garni Gorge, Armenia).

Figure 4.5. A tessellated rock shelf on the Tasmanian coast.

Figure 4.6. Tessellated features in the sandstone of White Pocket, on the border of Arizona and Utah.

An example is basalt columns with polygonal cross sections, often hexagonal, that result from the cooling of lava flows (Figure 4.7).

Divisions in plants and animals

Tessellations in nature are often used to form a barrier of some sort, and these barriers are generally constructed of smaller building blocks. One example is the protective layer of scales on lizards (Figure 4.8), snakes, and fish. Another type of barrier that forms a tessellation is the web of a spider (Figure 4.9). A turtle's shell is divided into sections that are often hexagonal. Tree bark is often divided into pieces that can be modeled as a tessellation on a cylinder (Figure 4.10).

The division of a layer into smaller units can facilitate biological processes, as in a layer of cells in a plant. The veins of a leaf, which facilitate the distribution of liquid, can divide it up into a tessellation, as shown in Figure 4.11. A similar division is seen in the wings

of insects such as dragonflies (Figures 4.12 and 4.13), though the primary function of the fluid, in this case, may be providing a sturdy scaffolding for a wing. A bee's honeycomb (Figure 4.14) is made up of hexagonal cells that closely approximate a regular tessellation of hexagons, allowing larvae to be stored in a compact fashion as they're cared for. Other cellular forms, such as that shown in Figure 4.15, can evolve as plants grow. Additional examples of tessellations in the natural world can be found in the book *Tessellations in Our World* [Fathauer 2014].

Figure 4.8. Lizard, near Tucson, Arizona.

Figure 4.9. Spider web, Phoenix, Arizona, and rotationally symmetric tessellation of trapezoids that approximates its structure.

Figure 4.10. The bark of a persimmon tree in Pennsylvania.

Figure 4.12. Dragonfly, Tokyo, Japan. The region of the right rear wing outlined in red was used to generate the tessellation of Figure 4.13.

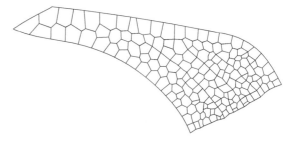

Figure 4.13. Drawing of the veins for the portion of the dragonfly wing outlined in red in Figure 4.12.

Figure 4.11. Veins in a maple leaf, Canadian Rockies.

Figure 4.14. Bee's honeycomb (MathLove museum, South Korea).

Figure 4.15. Fruit of an Osage orange tree in Pennsylvania.

Coloration in animals

Patches of color on animals, often used for camouflaging purposes, often form tessellations of sorts. Examples include the brown patches on a giraffe (Figure 4.16), the stripes on a zebra, and the color patterns on reptiles such as rattlesnakes and lizards (Figure 4.17). Patterns that could be modeled as tessellations are also observed in the wings of butterflies and moths.

Voronoi tessellations

Mathematical modeling can aid our understanding of biological processes. For example, the color patches on a giraffe's coat can be modeled as a Voronoi tessellation, as illustrated in Figure 4.18. A **Voronoi** *tessellation is created by drawing perpendicular lines midway between the lines joining the neighboring points in a collection of points and using those lines to define tiles.* Note that this process is related to dual tessellations, in which edges of a new tiling are drawn connecting centers of tiles in a starting tessellation. The fact that a giraffe's coloring closely resembles a Voronoi tessellation could support or contradict different models for how such pigmentation developed.

Figure 4.16. The patches of dark color on a giraffe can be thought of as tiles in a tessellation, though the lighter dividing lines are relatively wide (Sydney Zoo, Australia).

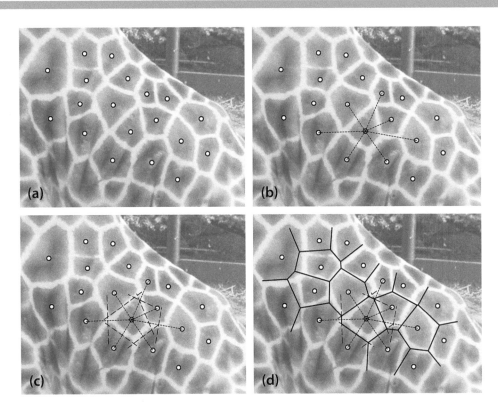

Figure 4.18. (a) Points located by eye in the center of the dark patches on a giraffe. (b) Dotted lines connecting one point to several nearby neighbors. (c) Dashed lines that are perpendicular bisectors to the dotted lines from (b). (d) Edges of a tessellation formed by the set of perpendicular bisectors for the collection of points from (a).

Activity 4.1. Modeling natural tessellations using geometric tessellations

Materials: Copies of Worksheet 4.1.

Objective: Learn to see geometric underpinnings of less regular, natural tessellations.

Vocabulary: Mathematical modeling, geometric tessellation, natural tessellation.

Specific Common Core State Standard for Mathematics addressed by the activity
    Use geometric shapes, their measures, and their properties to describe objects. (G-MG 1)

Activity Sequence
1. Write the vocabulary terms on the board and discuss the meaning of each one.
2. Pass out the worksheet.
3. Have the students draw tessellations as instructed on the worksheet.
4. Ask the students to share some of their designs.

Discussion Questions
1. Which one did you find more difficult, drawing the more naturalistic tessellation or simplifying it into a more regular geometric tessellation? Why?
2. Why do you think tessellations occur in nature?
3. What are some of the functions that tessellations accomplish in nature?
4. Can you think of some additional examples of tessellations in nature?

Worksheet 4.1. Modeling natural tessellations using geometric tessellations

For each real-world object, first, draw a portion as a tessellation of tiles that are close to the photograph in shape and then as a more geometrically regular tessellation, as shown in the example.

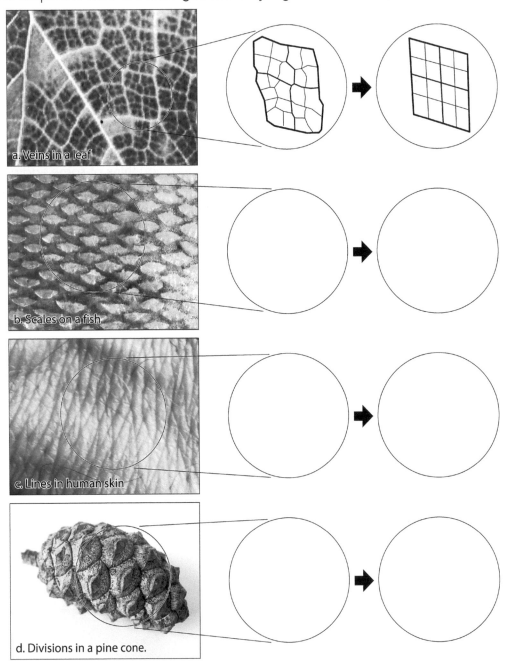

a. Veins in a leaf

b. Scales on a fish

c. Lines in human skin

d. Divisions in a pine cone.

Activity 4.2.
Quantitative analysis of natural tessellations

Materials: Copies of Worksheet 4.2.

Objective: Learn to analyze tessellations in nature quantitatively.

Vocabulary: Patch of tiles, Voronoi tessellation.

Activity Sequence

1. Write the vocabulary terms on the board and discuss the meaning of each one.
2. Pass out the worksheet.
3. Have the students create plots on the worksheet by counting the number of sides in the tiles for each patch of tiles.
4. Ask the students to share their results.

Discussion Questions

1. What did you find for the distribution of polygon types for the dragonfly's wing and the giraffe's spots?
2. How do the two differ?
3. Do you have any idea why there are differences?
4. If you wanted to carry these explorations further, what might be your next step?

The use of a Voronoi tessellation to model a giraffe's spots was described in the chapter. A 2018 article in Science News *describes a mathematical model for the divisions in the wings of dragonflies [Conover 2018]. That model also makes use of Voronoi tessellations. A key difference compared to the giraffe case is the presence of veins in a dragonfly's wing.*

Worksheet 4.2. Modeling natural tessellations using geometric tessellations

In the drawings of the dragonfly's wing and giraffe's spots, count the number of tiles with different numbers of sides. Plot the results on the axes provided, using a different color for the two animals.

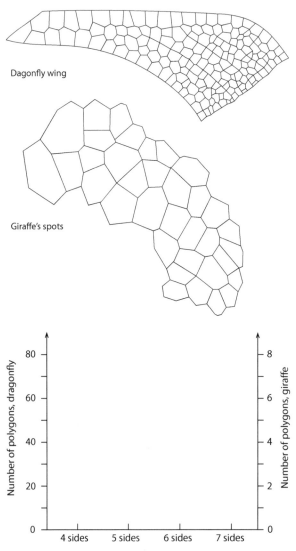

Dagonfly wing

Giraffe's spots

Chapter 5

Decorative and Utilitarian Tessellations

People of all cultures have used tessellations for the decoration of cloth, pots, floors, masks, and even their own bodies. In these applications, tessellations are used to make surfaces more beautiful, more distinct, or more interesting, though utilitarian aspects can be integral to the design. In addition, tessellations are widely used primarily for utilitarian, or practical, purposes. Artisans, designers, engineers, and architects often base designs on the three regular tessellations by equilateral triangles, squares, and hexagons, though less common tessellations do show up in many contexts. Some different types of uses are described in the following sections.

Tiling

Tiles in the traditional sense, made of ceramic or other materials, can be laid in attractive patterns, and individual tiles can be decorated with a tessellation as well. There are cultural and regional variations in the traditional use of tiles as regards material, style, and how widespread tiling is employed. In Portugal, for example, tiling is widely used on the exterior of buildings, including houses. In many cases, the square tiles are decorated with symmetric designs (Figure 5.1). Sophisticated and complex tilings are often seen in Morocco and other Islamic countries, particularly in the

Figure 5.1. House tiles in Porto, Portugal.

Figure 5.2. Tiled fountain in Rabat, Morocco.

Figure 5.3. Floor tiling in a doorway in London, England.

decoration of mosques, fountains, and palaces (Figure 5.2). Tiled floors are common in most parts of the world (Figure 5.3).

Building blocks and coverings

It is often the case that dividing up a large structure into smaller building blocks makes it easier or less expensive to fabricate that structure. The stacking of bricks and other building blocks with or without mortar seems likely to be one of the earliest uses of tessellations, with examples in ancient structures around the world (Figures 5.4 and 5.5). A wide variety of brick patterns can be observed in modern cities, particularly in Europe (Figure 5.6).

Small colored rocks are used to create larger street and sidewalk coverings that can be tessellated for aesthetic effect (Figure 5.7), and roofs are commonly covered with smaller shingles and tiles (Figure 5.8). Tessellations formed by sidewalk pavers are common in many countries, but not common in others like the United States. The capital of Armenia has examples of an unusually wide variety of tessellated pavers (Figure 5.9).

Tessellations can also be applied to structures for the sole purpose of making them more beautiful or distinct. Examples include the exterior of buildings (Figure 5.10) and utility covers. Considerable effort is put into creating beautiful utility cover designs in Japan, both tessellated (Figure 5.11) and otherwise.

Figure 5.4. Bricks and stone blocks were widely used by the Romans, as in the Pont du Gard aqueduct in the South of France.

Figure 5.7. Street covering in Funchal, Madiera Island, Portugal.

Figure 5.5. The Great Wall of China is constructed of bricks and other types of blocks.

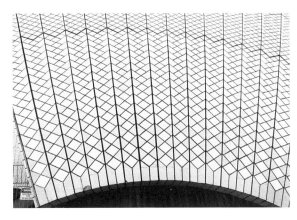

Figure 5.8. Tiles on the roof of the Sydney Opera House in Australia.

Figure 5.6. Three brick patterns in the Netherlands.

Figure 5.9. Tessellated pavers photographed in Yerevan, Armenia.

Figure 5.10. Decorative uses of tessellations on two buildings in Beijing, China, that are unrelated to structural function.

Permeable barriers

There are many situations which call for a structure that will act as a barrier to one or more things while allowing other things to pass through. Nets can provide barriers to people, fish, or balls while allowing light, air, and water to pass through. Examples include a soccer goal, a safety net around a play area (Figure 5.12), and a fisherman's net.

Fences can similarly act as barriers to people and animals while allowing light, air, and water through. The same is true of metal gratings over storm drains (Figure 5.13) and metal bars over windows (Figure 5.14).

Large windows are often divided into smaller panes of glass, blocking air and water while allowing light through (Figure 5.15). These can provide dramatic decorative effects, as evidenced by stained glass windows in churches. In more modern structures, tessellated glass panels can enclose large volumes, as shown in Figure 5.16.

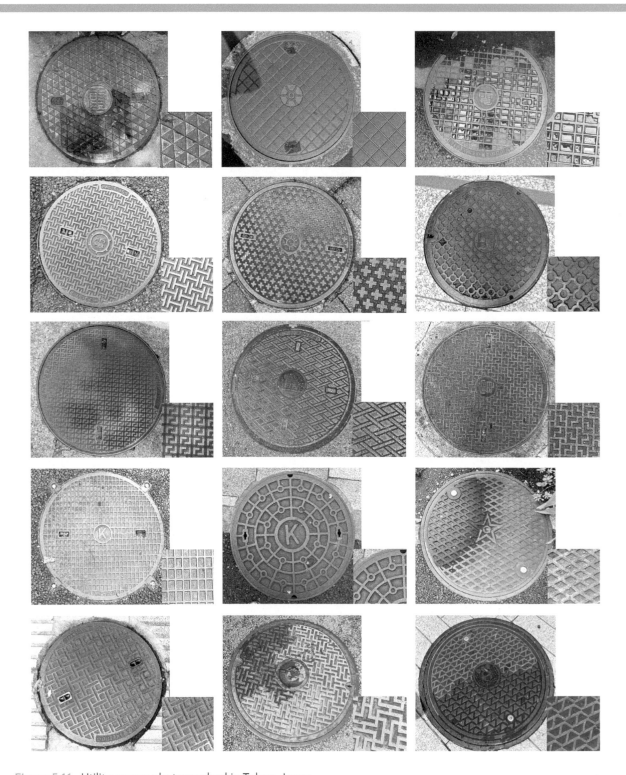

Figure 5.11. Utility covers photographed in Tokyo, Japan.

Figure 5.12. A net keeping children from falling out of a play area.

Figure 5.13. Metal grating over a storm drain in Seoul, South Korea.

Figure 5.14. Metal bars over the window of a mosque in Istanbul, Turkey.

Figure 5.15. Decorative window pane design in Kalmar Castle, Sweden.

Figure 5.16. Panes of glass in the covering of a courtyard in the British Museum, London, England.

Other divisions

There are other practical uses of tessellations. For example, farmland is divided into fields with different crops or different owners, as shown in Figure 5.17. Cloisonné is a technique in which metal divisions separate different colors of enamel to allow the creation of beautiful objects such as vases (Figure 5.18). The tread of automobile tires is broken into "tiles" to provide better traction, as shown by the imprint of Figure 5.19.

Figure 5.19. Tire track outside Tuba City, Arizona.

Fiber arts

Clothing is woven from fine strands to allow air to move through it while providing warmth and privacy. Quilts, which have a rich geometric tradition, are made from small pieces of cloth that traditionally were left over from the making of clothes (Figure 5.20). Woven baskets can hold a variety of foods and other items while allowing air and moisture to pass

Figure 5.17. Farmland outside Tokyo, Japan.

Figure 5.18. Cloisonné vase photographed at a factory showroom outside Beijing, China.

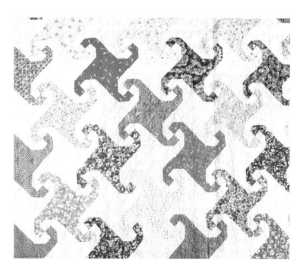

Figure 5.20. Quilt made in Oklahoma in the mid-20th century.

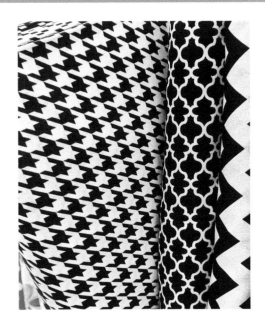

Figure 5.21. Tessellated designs on bolts of cloth.

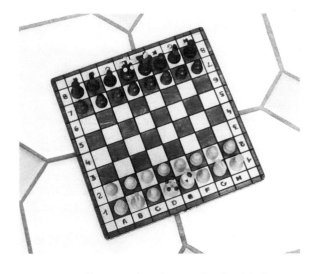

Figure 5.22. Chess is played on an 8 × 8 grid of squares.

through them to varying degrees. Lace is largely decorative, and traditional lace patterns exhibit a variety of geometric patterns. Cloth can be either woven into tessellations using different thread colors or printed with tessellations, as shown in Figure 5.21.

Games and puzzles

Tessellations have also been used widely in games and puzzles. Common examples include the squares in a chessboard or checkerboard (Figure 5.22), the divisions of a bingo card, a shuffleboard court, a dart board, and the division of an image into many small pieces in a jigsaw puzzle. There are also many puzzles in which tiles are used to form tessellations, as in Figure 5.23.

Figure 5.23. Wood tiling puzzle by Kevin Lee and Robert Fathauer, based on a Penrose tiling.

Islamic art and architecture

Islamic art has an ancient and rich tradition in the use of geometric tessellations. Beautiful examples can be seen around the world, particularly in mosques and other buildings in traditionally Muslim countries (Figure 5.24). Star polygons often play a prominent role in Islamic designs, and the edges of the tiles are sometimes replaced with interwoven latticework. Brian Wichmann and Tony Lee have created an online database of Islamic tilings [Wichman 2019], and Jay Bonner has recently written an in-depth book on the topic [Bonner 2017].

Figure 5.24. Five contemporary examples of the architectural use of Islamic art in Kuala Lumpur, Malaysia.

Spherical tessellations

Examples of spherical tessellations in the real world are not as common as planar tessellations, but they can be seen in a variety of settings. One practical application is the sewing together of panels of leather or cloth to form balls, the classic example being the spherical truncated icosahedron used for the traditional black-and-white soccer ball.

There are also sculptural examples, such as the sphere under the foot of the male lion in the Forbidden City, symbolizing power (Figure 5.25). Globes are typically marked with a grid that can be thought of as a spherical tessellation. A similar grid was employed in

Figure 5.26. Spherical cage supporting candles in the Stockholm Cathedral, Sweden.

Figure 5.25. Gilded lion in the Forbidden City, Beijing, China, with a close-up of the tessellated sphere under his right foot.

the construction of the spherical cage shown in Figure 5.26. Spherical tessellations are often featured in the Japanese art of Temari, in which intricate designs are embroidered onto spheres (Figure 5.27).

Figure 5.27. Temari ball with embroidered spherical tessellation. Image courtesy of Mrs. Kiyoko Urata. Photograph by Prof. Koji Miyazaki.

Activity 5.1. Building with tessellations

Materials: Copies of Worksheet 5.1.

Objective: Learn how tessellations can be used to add beauty to utilitarian structures.

Vocabulary: Utilitarian.

Activity Sequence

1. Write the vocabulary term on the board and discuss its meaning.
2. Pass out the worksheet.
3. Have the students draw tessellations as instructed on the worksheet.
4. Ask the students to share some of their designs.

Discussion Questions

1. Which of your three designs is your favorite?
2. Why do you like it?
3. What features do you think make for a good design?

Worksheet 5.1. Building with tessellations

Imagine you have two different types of bricks with which to pave a driveway, one square and one rectangular, as shown. Using the grid provided, design three different tessellations for the driveway that use both types of bricks. Try to make interesting and attractive designs.

Chapter 6

Polyforms and Reptiles

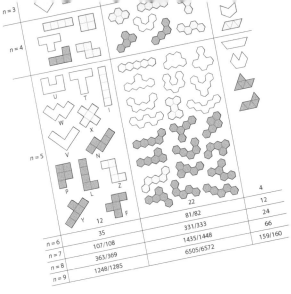

A *polyform* *is a polygon created by joining simple polygons in an edge-to-edge fashion.* The most common polyforms are based on regular tessellations. Polyominoes are formed by joining squares, polyhexes by joining regular hexagons, and polyiamonds by joining equilateral triangles. There are other polyforms as well, based on shapes such as the isosceles right triangle, but these are much less popular and won't be considered here. Polyforms can be thought of as supertiles, more complex tiles made by joining multiple copies of simple tiles. In building polyforms, one can envision putting tiles together or, equivalently, coloring adjacent spaces in a grid of the appropriate sort.

Solomon Golomb played a major role in developing polyominoes as a mathematical recreation [Golomb 1994]. A number of commercial puzzles and games have been developed based on polyforms. A particularly wide selection of plastic polyform-based puzzles is available from Kadon Enterprises [Gamepuzzle 2019]. In the popular computer game Tetris, falling polyominoes made up of four squares must be fit together in a tiled fashion.

Properties of polyforms

Polyforms are categorized primarily by the number of units n comprising them, starting with $n = 1$. An $n = 1$ polyomino is simply a square, but called in this context a monomino, analogous to the classic domino made up of two squares. An $n = 1$ polyhex is called a 1-hex or monohex, followed by the 2-hex or dihex. An $n = 1$ polyiamond is called a 1-iamond or moniamond, followed by the 2-iamond or diamond. There is only one way to join two squares, hexagons, or equilateral triangles edge-to-edge, so there is a single $n = 2$ polyform of each type.

With $n = 3$, things start to get more interesting. As seen in Figure 6.1, there are two trominoes, three 3-hexes (trihexes), and a single 3-iamond (triamond). Adding another unit, there are five tetrominoes, seven 4-hexes, and three 4-iamonds. At this point, it is seen that polyforms can possess a variety of symmetries. Of particular interest from the standpoint of recreational mathematics is whether or not a given polyform possesses mirror symmetry. If it does not, then flipping a tile over changes it. In the enumeration of polyforms, mirror variants are not generally considered to be distinct; e.g., there are five different tetrominoes rather than seven. In Figure 6.1, mirror symmetric polyforms are oriented with a vertical line of reflection, and they are in each box more or less above polyforms that lack mirror symmetry. The hexagon- and triangle-based polyforms are sometimes referred to using the same word prefixes as the polyominoes, but numbers will be used here.

The polyforms in Figure 6.1 are colored to indicate the main features of their symmetries. Polyforms with no symmetry are colored red; polyforms with mirror symmetry only are colored yellow; polyforms with rotational symmetry only are colored blue; and polyforms with both mirror and rotational symmetry are colored green. Note that polyominoes can possess one, two, or four lines of mirror symmetry and two- or four-fold rotational symmetry. Polyhexes can possess one, two, three, or six lines of mirror symmetry and two-, three-, or six-fold rotational symmetry. Polyiamonds can possess one, two, or three lines of mirror symmetry and two-, three-, or six-fold rotational symmetry. Try to find examples of these different symmetries in Figure 6.1.

When a fifth unit is added, the number of possibilities becomes relatively large for polyominoes and polyhexes. The 12 pentominoes are the most popular set of polyforms and have been used extensively in K–12 education. For ease of reference, they've been given names based on their resemblance to the letters of the alphabet, as shown in Figure 6.1. As additional units are added, the difference in the number of possibilities for polyominoes, polyhexes, and polyiamonds becomes increasingly striking. For example, with nine units there are over 40 times as many polyhexes as polyiamonds. With larger numbers of units, it becomes possible to create polyforms that contain holes. In Figure 6.1, the counts that have two numbers separated by a slash are the numbers without and with holes. Since tiles are not normally allowed to have holes, the first number is more relevant for the purposes of this book.

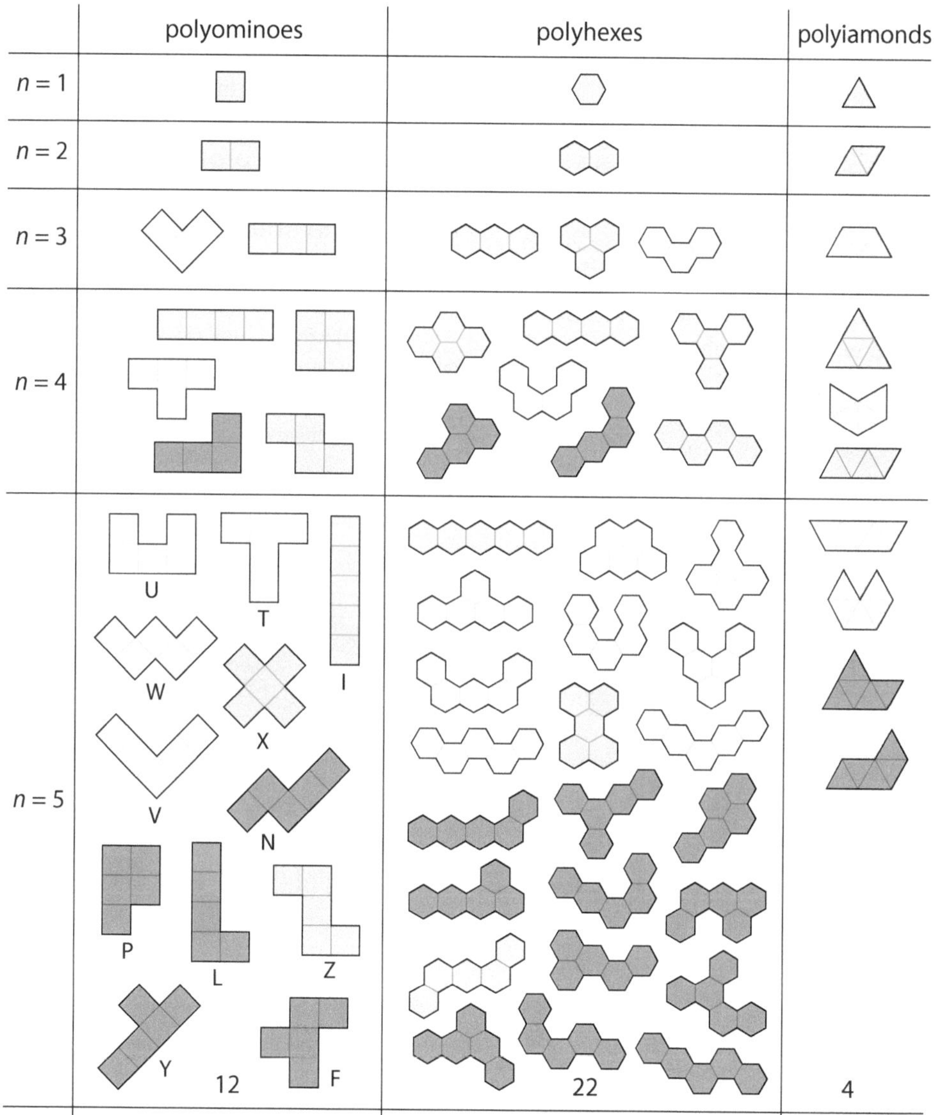

	polyominoes	polyhexes	polyiamonds
n = 1			
n = 2			
n = 3			
n = 4			
n = 5	12	22	4
n = 6	35	81/82	12
n = 7	107/108	331/333	24
n = 8	363/369	1435/1448	66
n = 9	1248/1285	6505/6572	159/160

Figure 6.1. Polyominoes, polyhexes, and polyiamonds made up of one to five units, along with the number of possibilities for six to nine units. Red denotes no symmetry, yellow mirror symmetry only, blue rotational symmetry only, and green both reflection and rotation.

Tessellations of polyforms

Tessellations using polyforms have been intensively studied, particularly the problem of which polyforms will tessellate, and how, using only copies of themselves [Clarke 2019, Grünbaum 1987, Myers 2019, Rhoads 2005]. It turns out all of the polyforms shown in Figure 6.1 will tessellate with themselves. For polyominoes, the first non-tilers are 3 of the 108 heptominoes. For polyhexes, the first non-tilers are 4 of the 81 6-hexes. For polyiamonds, the first non-tiler is 1 of the 24 7-iamonds. Interestingly, all 66 of the 8-iamonds tile the plane.

Note that, unlike the square and hexagon cases, the regular tessellation of triangles on which polyiamonds are based has two triangle orientations. Any tessellation of polyiamonds, therefore, when broken down into constituent triangles, must have equal numbers of the two triangle orientations. As a consequence, there are no tilings of polyiamonds with the odd n using only translation.

Examples of tilings with differing numbers of polyform orientations and different symmetries, made up of seven or eight units, are shown in Figures 6.2–6.4. In these figures, all tiles with the same orientation were given the same color. Well over 10,000 tilings of this

Figure 6.2. Examples of heptominoes that tile with a single orientation, require two orientations to tile, and require four orientations to tile.

Figure 6.3. Examples of 7-hexes that tile with a single orientation, require two orientations to tile, and require three orientations to tile.

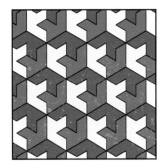

Figure 6.4. Examples of 8-iamonds that tile with a single orientation, require two orientations to tile, and require three orientations to tile.

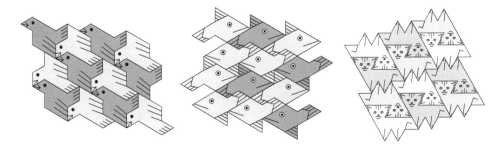

Figure 6.5. Simple Escheresque tessellations created by adding interior details to 11-iamonds that tile with two copies related by a 180° rotation.

sort are available for download from a website created by Joseph Myers [Myers 2019].

Some polyforms will inevitably suggest motifs to the mind, just as some clouds will. This is particularly true of polyiamonds. By adding a few interior details to such tiles, simple Escheresque tessellations can be created, as shown in Figure 6.5. Designs like this could be refined by reshaping edges as described in later chapters.

There are particular polyforms that have especially interesting tiling properties. An example is an 18-iamond known as the "loaded wheelbarrow", discovered by Roger Penrose to only tile the plane when used in 12 different orientations. Six are reflected relative to the other six, as shown in Figure 6.6.

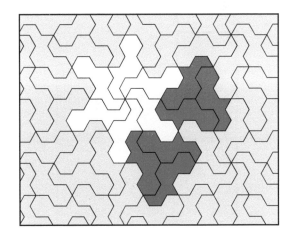

Figure 6.6. The "loaded wheelbarrow" 18-iamond tiling. The translation unit cell consists of the 12 yellow and red tiles, each of which is in a different orientation.

The translation and Conway criteria

Determining whether or not a prototile admits a tiling can be difficult. Fortunately, for many tilings, including most polyform tilings, two simple criteria will determine if it does. The first applies to prototiles that tile strictly by translation. *According to the **translation criterion**, a prototile tiles simply by translation, if and only if you can divide it into six segments, as shown in Figure 6.7, such that each of the pairs of segments AB–ED, BC–FE, and CD–AF are translations of each other. Both edges in one of these pairs may be empty.*

The second applies to prototiles that tile strictly by translation and 180° rotations. *According to the **Conway criterion**, a prototile tiles simply by translation and 180° rotations if you can divide it into six segments, as shown in Figure 6.7, such that the segments AB and ED are translations of each other and each of the remaining segments BC, CD, EF, and FA are centrosymmetric* [Wikipedia 2019]. *A **centrosymmetric** object is unchanged by a 180° rotation about its center.* Note that, unlike the translation criterion, the Conway criterion is not a necessary condition for tiling simply by translations and 180° rotations; i.e., a prototile could fail the condition and still tile in this manner. At least one of the segments of BC and CD and one of EF and FA must be nonempty. Segments AB and ED could be empty if at least three of the remaining four segments are nonempty.

Other recreations using polyforms

Polyforms lend themselves well to recreational mathematics and puzzles. The most common challenge is tiling rectangles with polyominoes. Such a tiling is a patch of tiles rather than a tessellation since it covers a finite area. However, rectangles can tile the plane, so, in a sense, solving a rectangle tiling is finding an infinite tessellation of polyominoes. One popular challenge is to arrange the 12 pentominoes in a rectangle using one of each type. The total area is 5 × 12 = 60 units, allowing in theory rectangles measuring 2 × 30, 3 × 10, 4 × 15, 5 × 12, and 6 × 10 to be constructed. The number of possible solutions for the five cases is 0, 2, 368, 1,010, and 2,339. Examples are shown in Figure 6.8.

Various figures can also be formed using polyhexes and polyiamonds. Stuart Coffin produced a wooden puzzled called "Snowflake" that consisted of one of each of the 3- and 4-hexes. Foam versions were later produced by Binary Arts and Tessellations. Some of the shapes that can be formed using the ten tiles are shown in Figure 6.9. Simple geometric shapes formed using one of each of the 12 6-iamonds are shown in Figure 6.10.

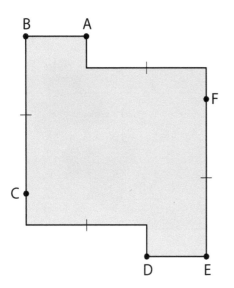

Figure 6.7. Labeling of a prototile for applying the translation and Conway criteria. The dots indicate points and the tick marks the midpoints of BC, CD, EF, and FA.

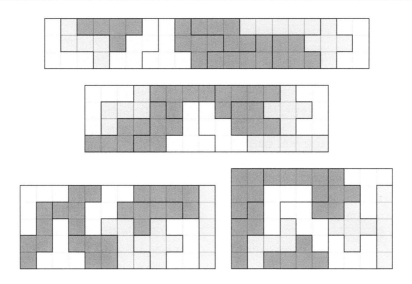

Figure 6.8. One each of the 12 pentominoes arranged in rectangles of different sizes.

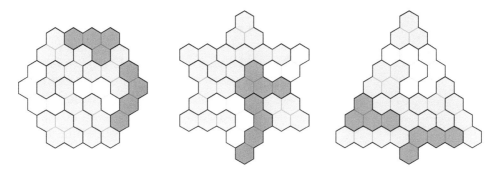

Figure 6.9. One each of the three 3-hexes and seven 4-hexes arranged to tile three different figures.

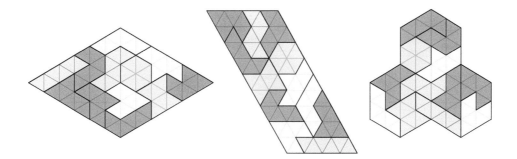

Figure 6.10. One each of the 12 6-iamonds arranged to tile three different figures.

Heesch number

A **corona** is a full layer of congruent copies of a tile T surrounding T. A corona must cover the corners as well as the edges, and the tiles in a corona cannot overlap or leave gaps. A second corona can then, for many tiles, be constructed similarly around the first corona. If T tiles the plane, coronas can be added indefinitely, provided coronas are chosen that allow the tiling to continue. More interesting is the case of tiles that don't tessellate. *The* **Heesch number** *is the maximum number of coronas that can be constructed around a non-tessellating tile.*

In 1968, Heinrich Heesch posed the problem of whether tiles with Heesch numbers of 1, 2, 3, exist…. Examples of tiles with a finite Heesch number greater than 1 weren't found until 1991, the first being U-shaped polyominoes. Robert Ammann discovered a notched hexagon tile that same year with Heesch number 3 (Figure 6.11). In 2004, Casey Mann described shapes called hexapillars with Heesch numbers 4 and 5 [Mann 2004].

A mathematical recreation using Heesch number and polyforms is to find the minimum and maximum number of tiles T in the corona of T. This is the basis for the game app Good Fences, written by Craig Kaplan and making use of his computer searches for tiles with large Heesch numbers based on polyforms and other shapes [Kaplan 2019]. Two examples are shown in Figure 6.12. Kaplan has identified polyiamonds with Heesch number 3 with as few as ten triangular units.

The current record for the tile with the largest Heesch number is still 5 as of early 2019. It has been conjectured that tiles exist for any positive integer Heesch number. This is one of the most interesting open questions in tiling theory.

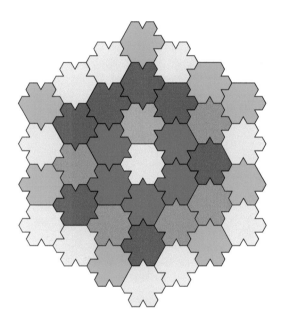

Figure 6.11. Amman's notched hexagon tile with Heesch number 3. Each corona is a different color, with the light shades being the same mirror variant as the center tile and the dark shades reflected.

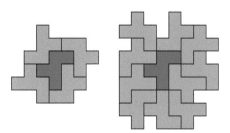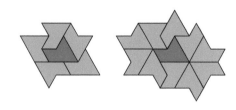

Figure 6.12. Minimum and maximum coronas for the F pentomino and a 5-iamond.

Reptiles

A ***self-replicating tile***, or "***reptile***" *for short, is a tile that can be tiled by smaller copies of itself.* Several polyominoes and polyiamonds have this property (Figure 6.13), but not polyhexes as a result of the manner in which hexagons tile. Note that a single square and a single equilateral triangle are both reptiles, but a single hexagon is not. The 6-iamond shown in Figure 6.13e, known as the "sphinx", is probably the most famous reptile.

By repeatedly subdividing self-replicating tiles into ever smaller copies of themselves, an arbitrarily large patch of tiles can be created, which is equivalent to being able to tile the plane; e.g., dividing the L tromino into four smaller trominoes four times yields the patch of tiles shown in Figure 6.14, with $4^4 = 256$ tiles. This process can be carried out indefinitely.

There are other polygons that are self-replicating, including a variety of triangles, as shown in Figure 6.15. A reptile that requires more than one size of a tile, such as the triangle in Figure 6.15b, is known as an irregular reptile, or "irreptile". More subtle examples include the "golden bee" tile discovered by Karl Scherer and a trapezoid discovered by Andrew Bayly (Figure 6.16) [Friedman 2018]. In addition, there are fractal reptiles, described in Chapter 9.

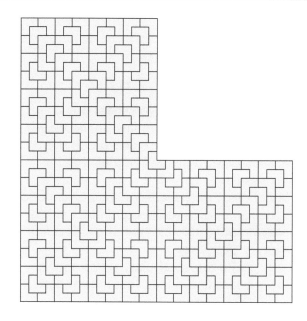

Figure 6.14. Four divisions of a tromino into four smaller trominoes results in a patch of 256 tiles.

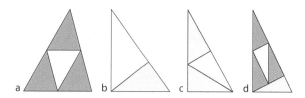

Figure 6.15. Tilings of triangles with similar triangles that (a) works for any triangle, (b) works for any right triangle, (c) only works for 30–60–90 triangles, and (d) only works for right triangles with short legs related by a factor of two.

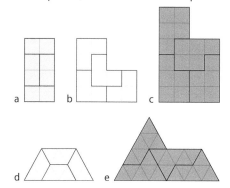

Figure 6.13. Reptiles based on (a) the domino, (b) the L tromino, (c) the P pentomino, (d) the 3-iamond, and (e) a 6-iamond.

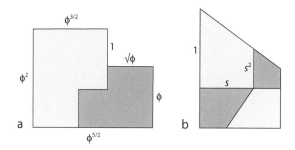

Figure 6.16. (a) Golden bee irreptile, where Φ is the golden number ≈ 1.618. (b) A trapezoid irreptile that works for a range of s, as shown here, $s = 1/\sqrt{2}$.

Activity 6.1.
Discovering and classifying polyforms

Materials: Copies of Worksheet 6.1.

Objective: Using a grid, find all the polyforms of a given type and group them according to the symmetries they possess.

Vocabulary: polyhex, tetrahex, polyiamond, hexiamond, rotational symmetry, mirror symmetry.

Activity Sequence

1. Write the vocabulary terms on the board and discuss the meaning of each one.
2. Pass out copies of the worksheet.
3. Using the grid, have the students find and draw the 7 distinct tetrahexes and 12 distinct hexiamonds. Be sure that they know mirrored variants are not considered distinct.
4. Have the students mark the symmetry that each polyform possesses and group them by symmetries by labeling them (A, B, C, …) and listing the number of tetrahexes and hexiamonds with each distinct set of symmetries (different combinations of rotational and mirror symmetries).

Discussion Questions

1. Have a couple of students share the tetrahexes or hexiamonds they found and see if the other students agree that they found all the distinct ones.
2. Have other students label the symmetries of different polyforms.
3. Discuss why certain symmetries are or are not allowed for tetrahexes and hexiamonds.
4. While there are only 12 distinct hexiamonds, there are more than 80 distinct hexahexes. Why do the students think there are a greater number of possibilities with polyhexes?

Worksheet 6.1. Discovering and classifying polyforms

There are 7 distinct tetrahexes and 12 distinct hexiamonds. One of each is shown on the grid. Try to find and draw all the remaining ones.

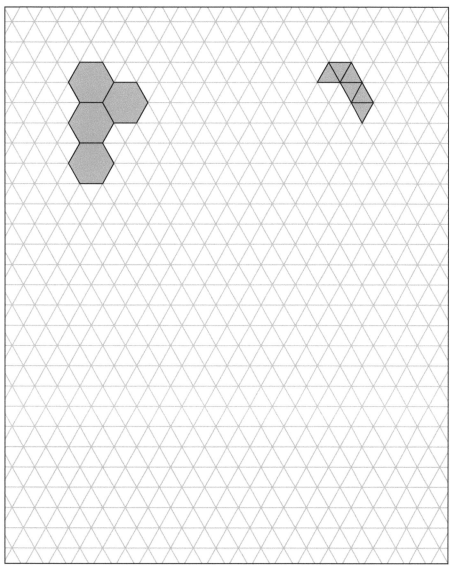

Chapter 7

Rosettes and Spirals

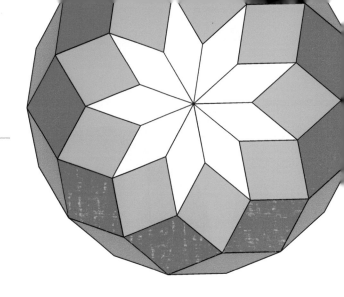

osettes and spirals, whether or not they are tessellations, are visually appealing due to their rotational symmetry and sense of movement. Tessellations that can be classified as rosettes or spirals are well-behaved tilings in many respects. However, rosettes are often not infinite in extent, and spiral tilings often contain a singular point at the center. This is the point near which tiles become infinitesimally small, and the density of tiles blows up. More precisely, *any disk, no matter how small, containing* **a singular point** *will meet an infinite number of tiles.* Singularity is another word for a singular point.

Rhombus rosettes

Broadly defined, **rosettes** are designs or objects that are flowerlike or possess a single point of rotational symmetry. They often contain lines of mirror symmetry through the rotation center as well. Rhombus-based tiling rosettes, with an *n*-fold rotational symmetry of the type shown in Figure 7.1, are relatively well known and can be constructed for any integer *n* greater than 2. Such rosettes are edge-to-edge, and every tile is a rhombus with the same edge length. While additional rhombi can be added at the perimeters, the basic rosettes are finite patches of tiles.

From Figure 7.1, it can be seen that there are some differences in rhombus rosettes for odd and even *n*. For odd *n*, each ring of the rosettes contains different rhombi, and the boundary of a rosette is a regular 2*n*-gon. For even *n*, each rhombus type appears in two rings, unless there are middle rings of squares, and the boundary of a rosette is a regular

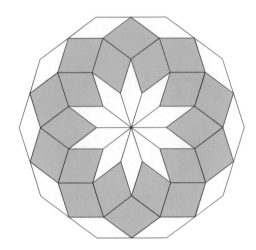

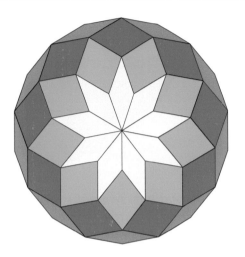

Figure 7.1. Rhombus rosettes for n = 9 and n = 10.

n-gon. Note that rhombus rosettes are sometimes referred to by the n-gon formed by the
boundary rather than the rotational symmetry.

For odd n, the number of rings is (n − 1)/2,
and the angles of the rhombi from the center

outward are (2π/n, (n − 2)π/n), (4π/n, (n −4)π/n),
... , ((n − 3)π/n, 3π/n), and ((n − 1)π/n, π/n). For
the rosette on the left side of Figure 7.1, n = 9,
giving angles of (2π/9, 7π/9), (4π/9, 5π/9), (6π/9,
3π/9), and (8π/9, π/9). Some additional examples

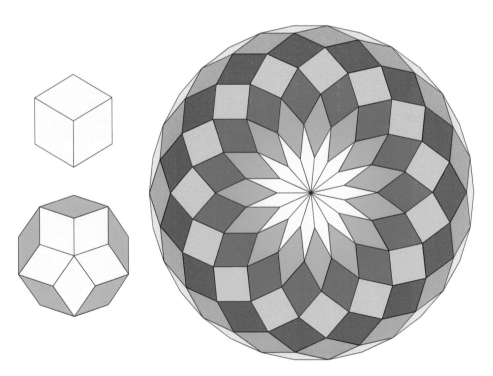

Figure 7.2. Rhombus rosettes for n = 3, 5, and 15.

are shown in Figure 7.2. Note that there are n lines of mirror symmetry, all of the same type.

For even n, the number of rings is $(n − 2)/2$. If n is a multiple of 4, the number of distinct rhombi is $n/4$, while it's $(n − 2)/4$ if n is not a multiple of 4. The angles of the rhombi from the center outward are $(2\pi/n, (n − 2)\pi/n)$, $(4\pi/n, (n − 4)\pi/n)$, ... , $((n − 4)\pi/n, 4\pi/n)$, and $((n − 2)\pi/n, 2\pi/n)$. If n is a multiple of 4, there are two rhombi of each type. If n is not a multiple of 4, there are two rhombi of each type except for the middle one, which is a square. Half-scale regular n-gons can be drawn from the center to the boundary, as shown with heavy lines in Figure 7.3. This is not possible for odd n. Note that there are again n lines of mirror symmetry but, in contrast to the odd case, of two different types.

A rhombus rosette can be thought of as a dissection of a regular n-gon into rhombi. For even n, the half-scale n-gons elucidate a second dissection. There are other ways to arrange these rhombi in a regular polygon that forfeit some or all of the symmetries [Schoen 2020]. For example, the ten rhombi in the $n = 5$ rosette have arrangements with a single mirror line and with no symmetry at all (Figure 7.4) [Wikipedia 2019]. Obviously, any tiling of regular n-gons can be converted into a tiling of rhombi by converting each n-gon into a rosette or into other arrangements of rhombi.

Rosettes serve as convenient templates for Escheresque tessellations. The first puzzle produced by my company, Tessellations, in 1993 was a set of Squids and Rays tiles based on the $n = 10$ rosette. Another puzzle with an insect theme was produced based on the $n = 14$ rosette, and a third with a reptile theme was produced based on the $n = 12$ rosette. Three of the motifs from that puzzle are shown in Figure 7.5. All these sets of tiles had the property with which they could fit together in numerous ways.

Figure 7.3. Rhombus rosettes for $n = 6$, 8, and 18. The heavy lines are half-scale regular n-gons.

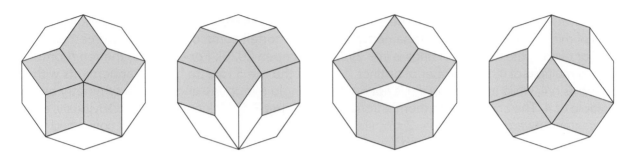

Figure 7.4. Four distinct tilings of a regular decagon with the same ten rhombi.

Figure 7.5. A patch of frog, horned lizard, and snake tiles based on an *n* = 12 rosette.

Other rosettes

In addition to rhombi, there are tiling rosettes based on other simple polygonal figures, including kites, triangles, darts, trapezoids, and hexagons. In this section, kite rosettes will be described that contain a singular point in the center. By marking the kites, rosettes with all the other shapes listed above can be formed.

In contrast to the rhombus case, there is only one prototile for these kite rosettes. For any *n* ≥ 1, a continuum of prototile shapes is possible, in contrast to rhombus rosettes [Fathauer 2018], in which there is only one version for each *n*. For *n* = 2, the kites are all

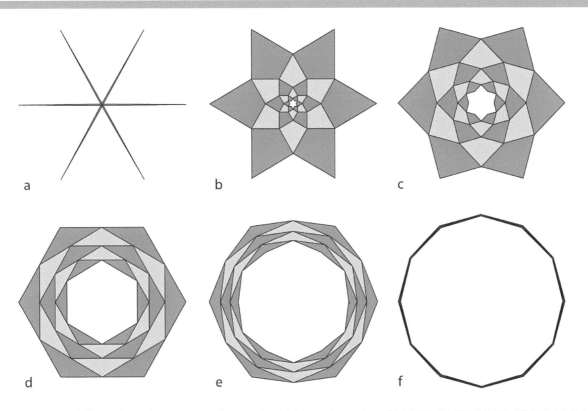

Figure 7.6. Six different kite-tiling rosettes for *n* = 6, with kite side angles of (a) 149°, (b) 90°, (c) 60°, (d) 30°, (e) 15°, and (f) 1°.

concave (darts), while for other *n*, they range from darts through an isosceles triangle to kites. In Figure 7.6 some of the kite rosettes for *n* = 6 are shown. Note that these kite rosettes are two-colorable. The scaling factor between adjacent rings of kites is given by the ratio of the long edge of the prototile to the short edge. The rosettes shown in the figures in this section can be considered annular patches of tilings that cover the infinite plane.

As with any tessellation, kite-tiling rosettes can serve as templates for Escheresque designs. An example is shown in Figure 7.7, where each kite in an eight-fold rosette has been turned into one flower and two leaves.

M.C. Escher made several prints with rotational symmetry about a singular point at the center. It doesn't appear that he was thinking in terms of kite-tiling rosettes as templates, though. His three "Path of Life" prints are based on smooth spirals, elucidated in looping lines down the centers of the fish and birds [Bool et al. 1982]. These can be fitted to kite-tiling rosettes; however, the angles of the kites are not round numbers, and there are two sets of decoration (to form the lifelike motifs) in alternate kites. "Development II" (1939) is an unusual case, based on a hexagonal grid with 24-fold symmetry (if colors are ignored). In the full print, a central hexagonal grid develops into lizards as the design progresses outward. This grid can be obtained from a kite-tiling rosette by applying simple marks to an appropriate prototile, as shown

Figure 7.7. An Escheresque design created by the author in 2011. The *n* = 8 kite-tiling-rosette template is shown on the right.

in Figure 7.8. These hexagons are not regular and deviate more from regular hexagons for smaller *n*.

In addition to the isosceles triangle tiling that lies at the transition from concave to convex kites for each *n*, triangle tilings can also be formed from kite rosettes by drawing lines between opposite corners of each tile. Lines can also be drawn to form trapezoid and pentagon tiling rosettes, as shown in Figure 7.9.

The visual appeal of these sorts of tilings can be extended to three dimensions by giving the tiles thickness [Fathauer 2018]. Two sculptural forms, with an architectural flavor, are shown in Figure 7.10. The example on the left has uniform tile thickness within each ring of a 15-fold kite rosette, while the example on the right varies thickness with rings as well as between them. A metal sculpture with some similarities to the left example was created by Simon Thomas [2020].

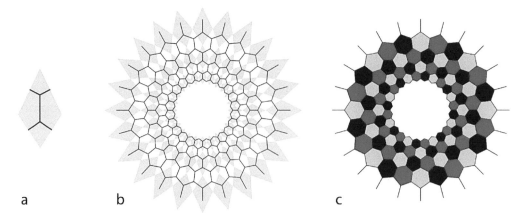

a   b   c

Figure 7.8. (a) An *n* = 24 kite with side angles of 120°, decorated with lines that generate a hexagonal grid (b) similar to that used in M.C. Escher's print "Development II". (c) A three coloring similar to that used by Escher.

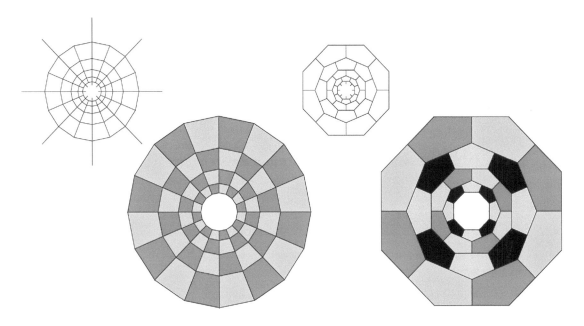

Figure 7.9. Examples of 16-fold trapezoid and 8-fold pentagon tiling rosettes obtained by marking the kites in an 8-fold kite-tiling rosette.

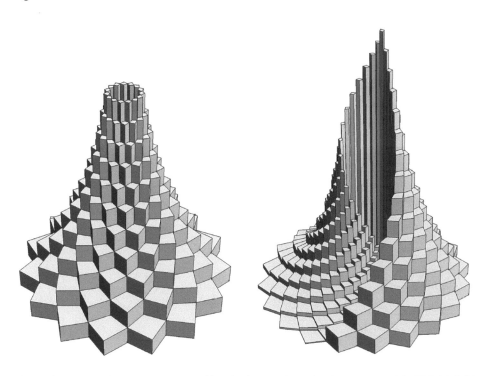

Figure 7.10. Three-dimensional structures created by thickening kite tiles in rosettes with 15-fold (left) and 20-fold (right) rotational symmetry.

Logarithmic spiral tessellations

A **spiral** *is a curve which emanates from a point, moving farther away as it revolves around the point* [Wikipedia 2019]. While it's difficult to precisely define what constitutes a **spiral tessellation**, a working definition *is a tessellation that emanates from a finite region that the tiles move farther away from as they approximately follow a spiral curve.* Spiral tessellations often contain a singular point at the center.

Most commonly encountered spiral tessellations can be described broadly as either Archimedean or logarithmic, the key difference being whether or not the tiles increase in size as distance from the center increases.

When describing spirals, it's more natural to use polar coordinates, where a point in the plane is specified by its distance r from the origin and its angle θ from the x-axis. In polar coordinates, *an **Archimedean spiral** has the equation* $r = \theta^{1/n}$, *where n is a constant that determines how tightly the spiral is wrapped* (Figure 7.11a). A **logarithmic spiral** *has the* *equation* $r = ae^{b\theta}$, *where a and b are constants* (Figure 7.11b) [Weisstein 2018].

The golden spiral and Fibonacci spiral are piecewise approximations to logarithmic spirals. These curves are formed by drawing quarter arcs of circles within squares, as shown in Figure 7.12. In the golden spiral tiling, each square is scaled by the **golden number**, an irrational number given by $(1 + \sqrt{5})/2 \approx 1.618$, relative to the next larger or smaller square. The Fibonacci spiral tiling is formed by squares of side lengths given by the Fibonacci sequence, in which each number is the sum of the two preceding numbers.

The golden number, commonly represented by the Greek letter phi, is more commonly referred to as the golden ratio and is also known as the golden mean, golden section, sacred cut, divine proportion, and other names. The term "golden ratio" implies a ratio of two objects, which is not how it will generally be used in this book. The number has a long history of being applied to objects that it doesn't really fit very well, such as the nautilus shell and the Acropolis. The term "golden number" is less fraught with these problematic associations.

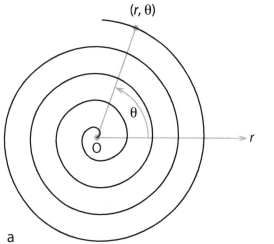
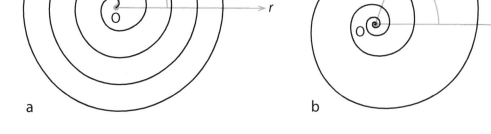

Figure 7.11. Archimedean (a) and logarithmic (b) spirals.

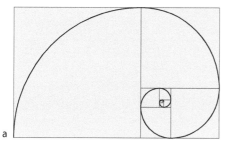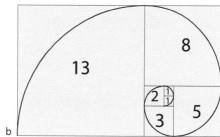

a   b

Figure 7.12. Golden and Fibonacci spiral tilings.

The golden spiral tiling of squares can be seen as a special case of the logarithmic rectangle tiling of Figure 7.13. For a golden spiral, the rectangular prototile is a square, so that $w = 1$.

Spiral tilings of equilateral triangles that are analogous to the square tilings of Figure 7.9 are shown in Figure 7.14. The scaling factor for Figure 7.10a is the plastic number, ≈ 1.325 [Sharp 2000]. In Figure 7.10b, integer triangle edge lengths are given by the Padovan sequence, in which each number is the sum of the two numbers preceding the immediately preceding number; i.e., a number is skipped

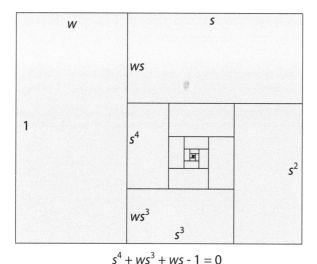

$$s^4 + ws^3 + ws - 1 = 0$$

Figure 7.13. A logarithmic spiral tiling of rectangles that becomes the golden square spiral when $w = 1$.

compared to the Fibonacci sequence. Spiral curves can be drawn on these tilings using 1/6 arcs of circles. The plastic number has also been used by Edmund Harriss to create a beautiful branched spiral based on a fractal tiling of squares [Bellos 2015].

There is another beautiful spiral tessellation in which each tile is an isosceles triangle with edge lengths related by the golden number, shown in Figure 7.15. The scaling factor between adjacent triangles is also the golden number, and a spiral curve can be drawn using 1/5 arcs of circles.

While the above spiral tilings have been known for some time, Cye Waldman pointed out a few years ago that the spiral of golden triangles is part of a family of spiral tilings of isosceles triangles that meet in a side-to-side fashion [Waldman 2016]. He and I independently realized that there are other families of spiral tilings of isosceles triangles. Examples of spiral tilings with three distinct types of matching of smaller-to-larger triangles are shown in Figure 7.16.

I extended these results by finding spiral tilings of isosceles triangles with more than one arm (Figure 7.17), spiral tilings of right triangles, and spiral tilings of scalene triangles (Figure 7.18) [Fathauer 2020].

Equations can be written for prototile angles and scaling factors (between successive triangles in the spiral) for each type of triangle and matching. For example, for an

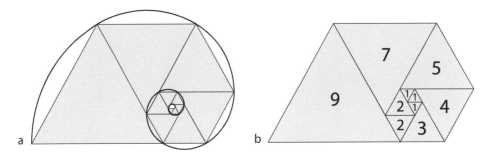

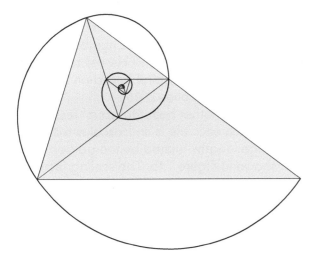

Figure 7.14. Plastic number and Padovan sequence spiral tilings.

Figure 7.15. Spiral tessellation of golden triangles.

m-armed spiral tiling of isosceles triangles meeting in a side-to-base (smaller-to-larger) fashion, the side angle α of the prototile is given by $(360°/m)/(n + 1)$ and the scaling factor s by $s^n + s = 2\cos(\alpha)$, where n is the

number of triangles in a single loop closing at a point. The two-armed tiling of Figure 7.17a is of particular interest for the fact that it is admitted by any isosceles triangle. All the other spiral tilings of triangles shown here are only admitted by discrete choices of angle.

The Fibonacci number is also observed in **phyllotaxis**, *the arrangement of features in plant growth*. Spirals occur naturally in the growth of a wide range of plants, from flower heads (the sunflower being a common example) to succulents to pine cones. For example, the wings of a pine cone are arranged in spirals, and the number of spirals in both clockwise and counter-clockwise directions is almost always a Fibonacci number. In Figure 7.19 a tessellation of pine-cone wings is drawn on a photograph of a pine cone that exhibits five clockwise and eight counter-clockwise spirals.

As seen in the phyllotaxis example above, spiral tessellations can have more than one arm. In addition, whether or not a tessellation

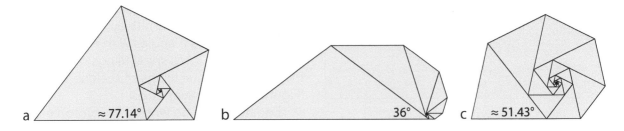

Figure 7.16. Examples of spiral tilings of isosceles triangles in which smaller triangles meet larger adjacent triangles (a) side to side, (b) base to side, and (c) side to base.

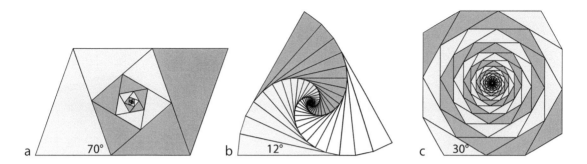

Figure 7.17. Examples of spiral tilings of isosceles triangles with more than one arm: (a) a double-armed spiral with base to side matching, (b) a triple-armed spiral with side to side matching, and (c) a quadruple-armed spiral with side to base matching.

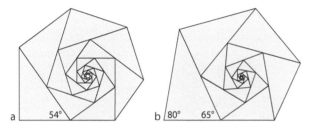

Figure 7.18. Examples of spiral tilings of (a) right triangles and (b) scalene triangles.

is perceived as being spiral can depend on how it's colored. For example, an *n*-fold kite rosette will be perceived as a spiral tiling with *n* arms if the tiles comprising each arm are colored the same as one another but differently from adjacent arms (Figure 7.20).

Logarithmic spiral tessellations in which the tiles increase at a uniform rate can be formed for a variety of simple polygons. Examples of single and double spirals using tiles that are nearly kite shaped are shown in Figure 7.18. **Double spiral** *is used here to mean a spiral that has two points at the perimeter where it can be extended.* **Double-armed spiral** *is used to mean a two-ended spiral, roughly similar in shape to the letter "S".* The uniform rate of increase doesn't allow strict kites, with only two edge lengths, to be used.

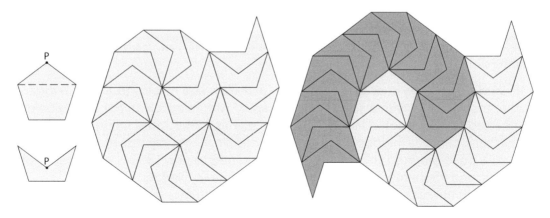

Figure 7.19. The construction of a reflected pentagon, with one-armed and two-armed spiral tilings based on it.

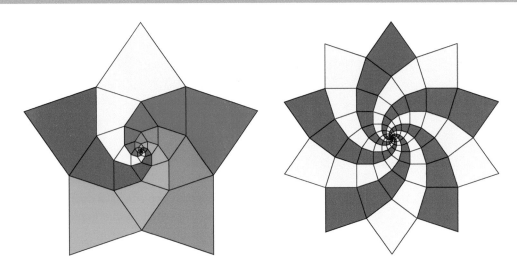

Figure 7.20. Two kite-tiling rosettes colored as spiral tilings.

Figure 7.21. Logarithmic spiral tilings with one and two arms using tiles that are nearly kite shaped. The tiles in the single spiral are shaded to enhance the structure of the spiral.

Spirals lend themselves to dramatic Escheresque tessellations, with a sense of motion imparted to the design. This is exemplified in Escher's prints "Whirlpools" (1957), which utilizes a double-armed spiral with two singular points, and "Sphere Surface with Fish" (1958), which utilizes a loxodromic spiral.

The similarity of the tiles in Figure 7.21 to kites allows an Escheresque tessellation that fits a kite rosette to also fit a logarithmic spiral with relatively minor distortions of the tile. This is illustrated in Figure 7.22, where a periodic manta ray tessellation (a) has been modified to fit a 13-fold kite rosette (b) and logarithmic tiling (c).

Logarithmic spiral tilings can be generated by mapping a periodic planar tiling to the complex plane using a simple exponential

Figure 7.22. A periodic manta ray tessellation (a) adapted to fit a kite rosette (b) and a logarithmic tiling (c). The geometric tile on which the manta ray is based is shown for all three in yellow.

function known as the anti-Mercator mapping [Kaplan 2019]. With a suitable choice of parameters, the mapping will convert a given planar tiling to a spiral tiling with various numbers of spiral arms in each direction. For example, the planar tessellation of three-lobed tiles shown in Figure 7.23a can be mapped to a spiral tiling with a differing number of arms (Figure 7.23b,c). Similarly, logarithmic spirals with two singular points can be generated by a Möbius transformation (Figure 7.23d). Tilings of the sort shown in Figure 7.23 can be generated interactively online using a tool developed by Craig Kaplan [Kaplan 2019].

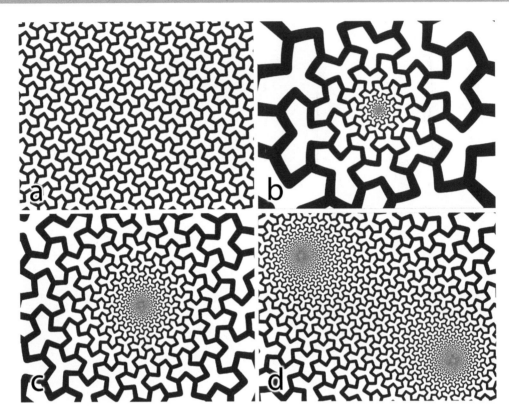

Figure 7.23. A planar tiling and three spiral tilings created by mapping it to the complex plane.

Archimedean spiral tessellations

All the spiral tilings shown so far are logarithmic, with tiles that increase in size as the spiral evolves outward from the center. The remaining examples will be more or less Archimedean in nature. In the logarithmic examples, all the tiles in a given spiral tiling are similar, but in the Archimedean examples below they're all congruent. Tilings of this general sort can be constructed with any number of arms [Stock and Wichmann 2000]. It's also possible to construct spiral tilings with more than one prototile, but single-prototile examples are generally of more interest.

Archimedean spiral tessellations can be constructed using squares and equilateral triangles. These can have a single point of rotational symmetry, with the arms spiraling outward endlessly. Using a regular tessellation of squares, one, two, or four spirals can originate from the center point, as shown in Figure 7.24. With a single spiral, delineating the spiral requires some sort of decoration or color progression.

Using squares and equilateral triangles, periodic tessellations are also possible with two-singular-point spirals or multi-armed spirals in which the number of segments in each spiral can be varied [Fathauer 2019]. These latter tilings are related to the three regular tessellations and have considerable visual appeal. Monohedral tessellations with tiles created by joining squares are shown in Figure 7.25. Triangle-based examples in

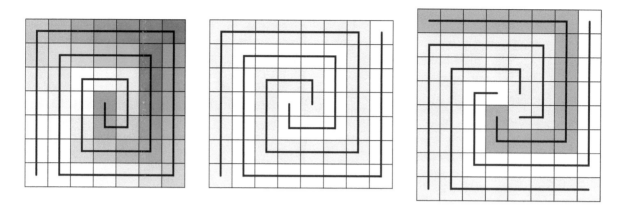

Figure 7.24. One, two, and four spirals that can be extended endlessly from a single rotation center using a regular tessellation of squares.

Figure 7.25. Tessellations of double-armed-spiral tiles (left) and four-armed-spiral tiles (right). The tiles are comprised of squares, a few of which are outlined on the left. The arms of the spirals turn by 90° between adjacent segments.

which adjacent arms turn by 120° are shown in Figure 7.26, and triangle-based examples in which adjacent arms turn by 60° are shown in Figure 7.27. The same sorts of spirals could be constructed from a regular hexagon tessellation, but the straight segments would be lumpy due to the natures of the tiles.

Another simple prototile that can be used to construct spiral tessellations is an isosceles triangle of the sort shown in Figure 7.28. In this example, the prototile is a 1/8 wedge of a regular octagon, but analogous tilings are allowed by a similar wedge of any 2n-gon [Grünbaum 1987]. A tiling with 8-fold rotational symmetry is shown in Figure 7.28a. Spiral tessellations with different numbers of arms can be obtained simply by shifting half the tiling, as shown in b and c.

Figure 7.26. Tessellations of double-armed-spiral tiles (left) and three-armed-spiral tiles (right). The tiles are comprised of equilateral triangles, a few of which are outlined on the left. The arms of the spirals turn by 120° between adjacent segments.

Figure 7.27. Tessellations of double-armed-spiral tiles (left) and six-armed-spiral tiles

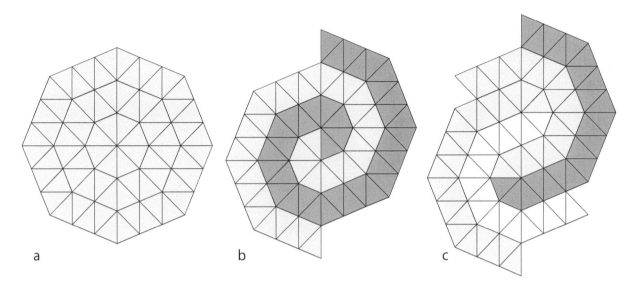

a   b   c

Figure 7.28. Three tilings with a prototile that is a 1/8 wedge of a regular octagon. Shifting half tiling (a) by different amounts creates spiral tilings with different numbers of arms (b). In (c) some additional tiles were added to elucidate the four arms.

Another type of prototile that lends itself well to spiral tessellations is a reflected regular *n*-gon. The pentagon version of such a tile is shown in Figure 7.29. Related spiral tilings based on prototiles obtained by joining tiles in rhombus rosettes have also been described [Gailiunas 2007]. A variety of periodic tessellations are also allowed by such prototiles.

Figure 7.29. A photograph of a pine cone with a spiral tessellation drawn on it. There are five clockwise (blue) and eight counter-clockwise (red) spirals of pine-cone wings.

The Voderberg tiling is another spiral tiling in this class that makes use of a more sophisticated prototile. The Voderberg's tile has the remarkable property that two tiles can completely surround a third, or even two additional tiles, as shown in Figure 7.30 [Grünbaum 1987].

Figure 7.30. Voderberg spiral tiling.

Activity 7.1. Exploring spiral tessellations

Materials: Copies of Worksheet 7.1, scissors, and colored pencils or other type of marker are optional.

Objective: Create spiral tessellations of different types from a regular tessellation and explore variations.

Vocabulary: Spiral, double spiral, regular tessellation, spiral tessellation.

Activity Sequence

1. Write the vocabulary terms on the board and discuss the meaning of each one.
2. Pass out two copies of the worksheet to each student, along with scissors and optional colored markers.
3. Have the students complete the drawing of the single spiral and then cut the worksheet along the indicated line.
4. Have the students set the two halves in their original positions, then shift the bottom half one square to the right and describe what they observe in terms of the number of spirals, single vs. double, and clockwise vs. counter-clockwise. Have them shift two squares and then three to the right, and then one, two, and three squares to the left from the starting position, describing what happens each time. Coloring each path a different color may make it easier to see what's happening.
5. Have the students complete the drawing of the double spiral on the second sheet and then cut the worksheet along the indicated line.
6. Have the students shift the paper and describe what they observe as in step 4 for the single spiral.

Discussion Question

1. What do you think would happen if the single spiral was shifted four squares to the left? Test your answer.
2. What do you think would happen if the double spiral was shifted four squares to the right? Test your answer.
3. In what sense are the various tessellations you looked at are spiral tessellations? (Marking squares with lines is just a simpler alternative to coloring.)
4. How do the top halves of the sheets compare for the single and double spirals? How about the bottom halves?

Worksheet 7.1. Exploring spiral tessellations

On one copy of the worksheet, ignore the short dashes and continue drawing the single spiral that starts at the black dot. Try to keep your line centered in the squares and draw until all the squares are filled. Then cut the sheet in two parts along the horizontal dashed line.

On the other copy of the worksheet, draw over the short dashes and continue the line in two directions to draw a double spiral centered at the black dot. Then cut the sheet in two parts along the horizontal dashed line.

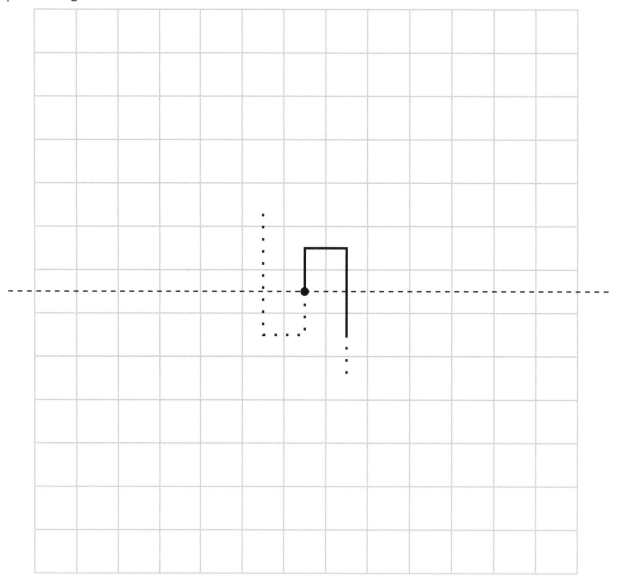

Chapter 8

Matching Rules, Aperiodic Tiles, and Substitution Tilings

Most of the examples of geometric tessellations presented to this point were constructed with unmarked tiles. The only rule for fitting tiles together was that, for the most part, the fittings must be edge-to-edge. Marking tiles to enforce more restrictive matching rules is also of interest, particularly when it comes to aperiodic tiles. When it comes to designing Escheresque tessellations, matching rules determine the sort of shapes that are possible.

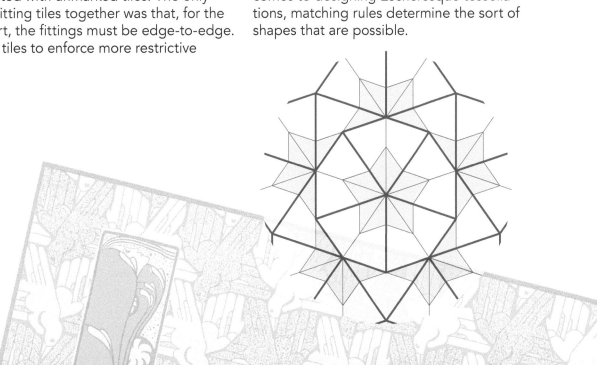

Matching rules and tiling

A matching rule is a law specifying how tiles in a tessellation can be fit together. Such rules can be indicated by graphics or by distortions of an edge. Three examples are shown in Figure 8.1.

The choice of matching rules completely changes the tilings admitted by a particular set of prototiles. For example, a pair of rhombi with angles that are multiples of π/5 are shown in Figure 8.2. The markings in Figure 8.2a only

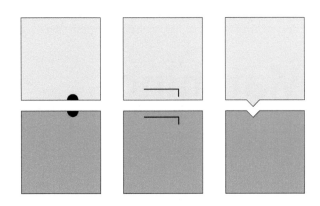

Figure 8.1. Three ways of marking the edges of two tiles to indicate they can fit together.

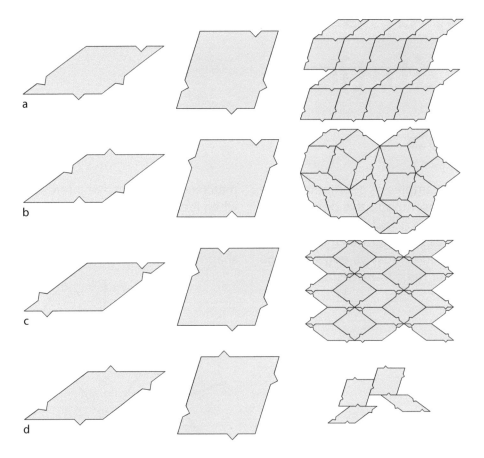

Figure 8.2. Four sets of matching rules for rhombi with angles 36°/144° and 72°/108°. (a) A combination that only tiles periodically, except for trivial variations. (b) A combination that only tiles non-periodically (the Penrose set P3). (c) A combination that tiles both periodically and non-periodically. (d) A combination that doesn't tile at all.

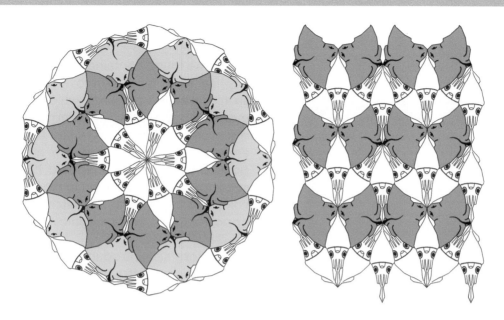

Figure 8.3. Two of the many tessellations allowed by squid and ray tiles based on the matching rules of Figure 8.2c.

allow a periodic tiling, except for the sort of trivial variations described below. The markings of Figure 8.2c force both tiles to have bilateral symmetry. This is convenient for lifelike motifs since most animals have approximate bilateral symmetry. This particular set of matching rules allows a wide variety of tessellations, some repeating, some with rotational symmetry, and some with no symmetry. This variety allowed me to use this template for my first tessellation puzzle, "Squids and Rays" (Figure 8.3), which was first produced in 1993.

The matching rules of Figure 8.2b remarkably only allow tessellations that cover the infinite mathematical plane without repeating. This famous aperiodic set will be described in more detail below. The matching rules of Figure 8.2d do not allow any tessellations, even if the tiles are allowed to fit in a non-edge-to-edge fashion.

Periodicity in tessellations

A periodic tiling is one possessing translational symmetry, while a non-periodic tiling is one that does not possess translational symmetry. An aperiodic set of tiles is one that only admits non-periodic tilings. The tiles of Figure 8.2a will tile periodically, as shown. One could imagine arranging tiles of this general sort in a non-periodic fashion as well. For example, the tiling would not repeat in the vertical direction if three rows of skinny rhombi were followed by one row of fat rhombi, by four rows of skinny rhombi, by one row of fat rhombi, etc. The number of rows could continue to progress as the digits of π increase, which never repeat. Mathematicians would call this a trivial variation of the periodic tiling, which isn't very interesting. In this

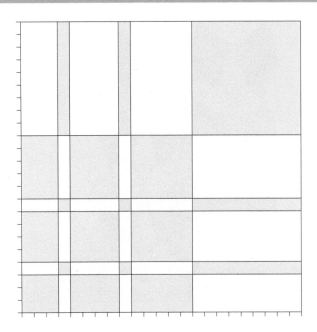

Figure 8.4. A patch of a trivial non-periodic tiling based on the digits of π.

repeating. It had previously been conjectured that no such set existed. However, the smallest set he found contained 104 tiles. In 1971, Raphael M. Robinson discovered a set of only six tiles with this property, which are basically modified squares. In 1973, Roger Penrose discovered a different set of six tiles which are related to five-pointed stars and pentagons [Grünbaum and Shephard 1987]. This set, known as P1, is shown in Figure 8.5 with colored circles indicating matching rules, as opposed to the edge modification originally used by Penrose.

Penrose also introduced a related set of tiles in which the straight lines are replaced by arcs of circles. This set, shown in Figure 8.6, is not equivalent to P1 but has been conjectured to also be aperiodic. By combining some of the tiles and adding graphics,

particular example, the tiling would still be periodic, as it would repeat in the horizontal direction. A non-periodic tiling based on this approach can readily be constructed using rectangles, as shown in Figure 8.4.

If you're bothered by the fact that the tiling of Figure 8.4 example only extends to infinity to the right and up, the rest of the planes can be tiled by mirroring vertically and horizontally. The resulting mirror symmetry does not allow translation. Similarly, a single point of rotational symmetry in an infinite tiling of rhombi or other shapes does not make a tiling periodic, as it still lacks translational symmetry.

Penrose tiles

In 1966, Robert Berger made the counterintuitive discovery of the first set of tiles that would admit infinitely many tilings of the plane, none of which are periodic or

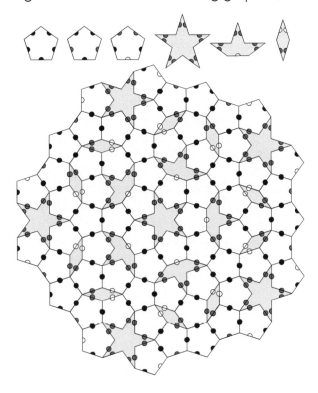

Figure 8.5. A patch formed by an aperiodic set of six tiles discovered by Roger Penrose.

I created a puzzle called "In the Garden" that allows both periodic and non-periodic tessellations.

In 1974, Penrose succeeded in discovering two closely related sets of only two tiles, the angles of which are multiples of π/5 (Figure 8.7). The "kites and darts" set was popularized by Martin Gardner in January, 1977, issue of *Scientific American*, which features a tiling of kites and darts on the cover [Gardner 1977]. Matching rules are indicated by circular arcs which must be continuous across boundaries between tiles. Non-edge-to-edge matchings are also not allowed. If the matching rules are followed, an infinite number of different tessellations can be constructed, all of which will be non-periodic.

Figure 8.6. Patches of a set of six tiles derived from the set P1 (left) and a tessellation puzzle called "In the Garden" (right).

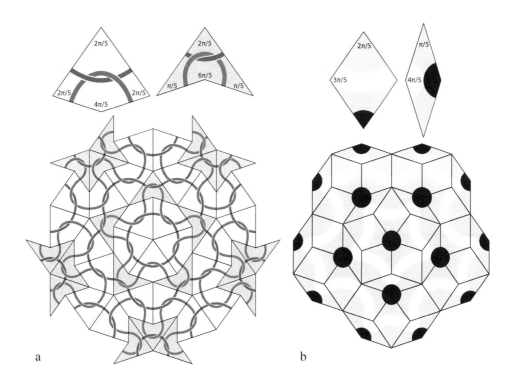

Figure 8.7. Two related sets of two prototiles that only tile non-periodically, along with a small patch of a representative tiling. (a) Penrose kite and dart and (b) Penrose rhombi.

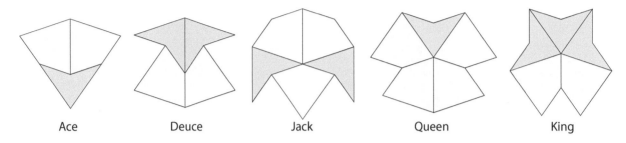

Figure 8.8. The five allowed vertex groups for Penrose kites and darts that are in addition to the sun and star groups.

One interesting property of Penrose tiles is how often the golden number (1.618 …) pops up. You'll notice that the edges of the kites and darts have two lengths, which can be referred to as "long" and "short". The long edge is exactly the golden number times the short edge in length. Furthermore, the area of a kite piece is exactly the golden number times the area of a dart piece. In any finite tessellation of kites and darts containing more than just a few pieces, the ratio of kites to darts is approximately equal to the golden number, and in any infinite tessellation, it is precisely the golden number.

One way to characterize a set of tiles is by looking at the allowed vertices. Recall that this was done in Chapter 2 for regular polygons. For Penrose tiles, there are seven allowed vertex types. One is the group of five kites at the center of Figure 8.7a. Another is the group of five darts at the center of Figure 8.7b. These are respectively known as the Sun and Star. The other five are shown in Figure 8.8. In every infinite tessellation of kites and darts, every one of the seven types of vertices occurs and so do an infinite number of times. By way of contrast, the number of distinct vertices allowed by the Squids and Rays tiles of Figure 8.3 is 15.

Further characterizing a set of tiles by vertex groups, the number of tiles forced into certain positions by the group can also be

examined. For example, the Star group forces ten kites in a ring around it, but there are choices after that. The "empire" of the group is the collection of forced pieces which must be in any tessellation surrounding that group. Some of these pieces immediately surround the group, and others are separated from it by unforced pieces. The Ace forces no more pieces, while there are only two pieces immediately forced by the Deuce. The King reigns, with the largest empire of all, immediately forcing 50 tiles. Many of the fascinating properties of Penrose tiles were first discovered by mathematician John Conway [Gardner 1977].

A key property of Penrose tiles is composition and decomposition. Decomposition is the process of dividing tiles into smaller tiles. Composition is the converse process. More precisely, a tiling T1 is obtained from the composition of a tiling T2 if each tile of T1 is a union of tiles of T2 [Grünbaum and Shephard 1987]. The terms "inflation" and "deflation" are also used in the literature, especially for Penrose tiles, but I find them ambiguous as to which term refers to which process. Kites and darts can be decomposed into kites and darts, and the same is true for the Penrose set P1 and the Penrose rhombi. To decompose a patch of kites and darts, imagine each dart is divided into a kite and two half darts and each kite into two kites and two half darts, as shown in Figure 8.9. The half darts are then

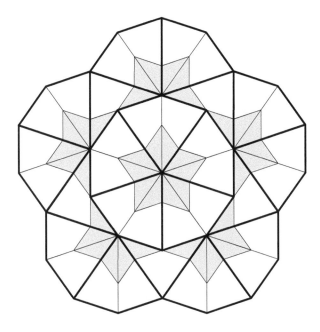

Figure 8.9. Decomposition applied to a patch of the infinite sun tiling, outlined in heavy black lines. The new and smaller kites and darts are colored and form a patch of the infinite star tiling.

merged to form smaller darts. As shown for a small patch of tiles in Figure 8.9, this process converts the Penrose tiling known as the infinite sun (also shown in Figure 8.7) into a tiling known as the infinite star. Decomposition can be carried out ad infinitum. A related modification of Penrose kites and darts is to cut each tile along its axis of symmetry, forming two types of golden isosceles triangles known as Robinson triangles. These facilitate the conversion of a kite and dart tiling into a rhombi tiling and vice versa.

The tessellation of Figure 8.10 is known as the Infinite Cartwheel. The regular decagon outlined with a heavy line is referred to as a cartwheel. Every point in every kite and dart tessellation lies inside or on the boundary of a cartwheel. The particular arrangement of kites and darts shown within the boundary of the cartwheel is the only legal tiling of this

shape. The spokes in the Infinite Cartwheel tessellation, composed of long (light blue) and short (light green) Bowties, are known as Conway worms. Every kite and dart tessellation contains an infinite number of arbitrarily long worms. Between the spokes are regions which get larger further away from the central cartwheel. These regions exhibit alternating and increasingly large portions of the infinite sun and star tessellations.

It's been proven that periodic tessellations can only contain rotational points with two-, three-, four-, or six-fold symmetry (see the plane symmetry groups in Chapter 2). In 1984, Daniel Shechtman et al. created shockwaves in the mathematics and physics community by publishing crystalline diffraction patterns with five-fold symmetry [Shechtman et al. 1984]. Because of the mathematical prohibition against five-fold symmetry in a lattice, it was widely assumed that crystals possessing such symmetry and therefore capable of producing such diffraction patterns could simply not exist.

The rhombic version of Penrose tiles played an important role in solving this dilemma. It was shown that quasi-periodic structures with regions possessing five-fold symmetry and sufficient long-range order, but not strict translational symmetry, could give rise to the sort of diffraction pattern observed. Such materials came to be known as quasicrystals. Three-dimensional Penrose tilings were constructed using two different rhombohedra (polyhedral with rhombic faces) [Nelson 1986] and were instrumental in understanding the crystalline structure of real-world quasicrystals. One aspect of this work was the discovery that Penrose rhombus and related non-periodic tilings could be obtained from projections of higher-dimensional lattices. In addition to five-fold, such projections could have other local rotational symmetries including seven-fold and nine-fold [Whittaker 1988].

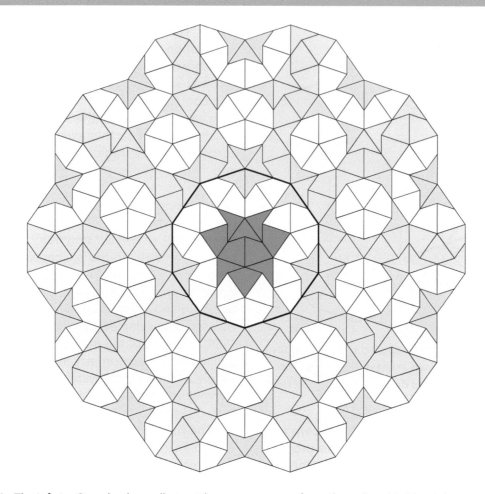

Figure 8.10. The Infinite Cartwheel tessellation. The center group of ten tiles colored in blue is known as Batman, and a regular decagon formed by the central 40 tiles is outlined in heavy black lines. The spokes of the cartwheel (Conway worms) are composed of long (light blue) and short (light green) Bowties.

Other aperiodic sets and substitution tilings

About the same time Penrose discovered his aperiodic sets, Robert Ammann discovered related sets of aperiodic tiles. Unlike Penrose, a well-known academic, Ammann was an American post office employee and amateur mathematician. For this reason, his work didn't receive as much notice as that of Penrose, especially early on. One of Ammann's sets, known as A2, is shown in Figure 8.11, with black ovals that enforce the matching rules. Mirroring is allowed in addition to rotation, and all the tile angles are right angles.

Unlike Penrose tiles, the edge lengths can be varied in the set A2. However, if the short and long edges of the smaller tile are related by the golden number, additional properties emerge. One of these is the ability to mark the matching rules with Ammann bars, a straight line that continues unbroken across a legal tiling. Ammann bars are shown in Figure 8.11, where either the bars or the ovals will enforce matching rules. Penrose tiles can also

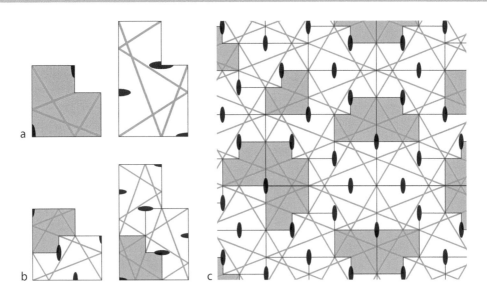

Figure 8.11. Ammann's aperiodic set A2. (a) The two prototiles. (b) Substitution rule for increasing the number of tiles. (c) A portion of a tiling.

be marked with Ammann bars [Grünbaum and Shephard 1987].

In Figure 8.11b, the decomposition rules for A2 are shown. Another word for decomposition in this context is "substitution", and mathematicians used the term "substitution tiling" to describe a tiling that can be constructed by iterative substitution according to such a set of rules. If an A2 tiling is constructed, from a single starting tile of either type, by iterative substitution, then markings enforcing matching rules are not needed. The substitution rules ensure proper matching and take the place of matching-rule markings.

Iterative substitution is a powerful method of constructing non-periodic tilings. It can even be applied to a tile or set of tiles that are not aperiodic. For example, the L tromino, or

"Chair", will admit a non-periodic tiling when the substitution rule of Figure 8.12 is used. However, this tile also admits periodic tilings. A related set of two tiles that is aperiodic is the Trilobite and Cross, discovered by Chaim Goodman-Strauss [Goodman-Strauss 1999].

Ammann's aperiodic set A5 is the best known of his sets. It scales as $1 + \sqrt{2}$ and has patches with local eight-fold symmetry. It's more commonly called the Ammann–Beenker tiling, recognizing F. Beenker's algebraic analysis of the tiling [Frettlöh et al. 2019]. Patches of tiles are shown in Figure 8.13, along with the substitution rule. There is an online database of substitution tilings by Dirk Frettlöh et al. that contains several of the tilings in this chapter and many more examples [Frettlöh et al. 2019].

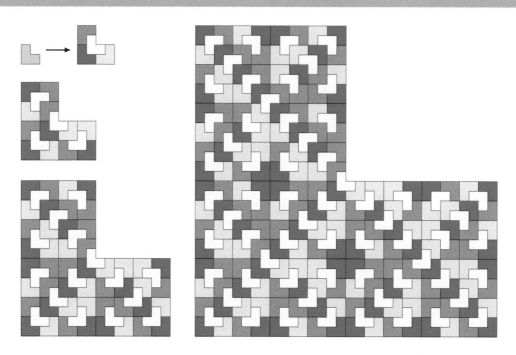

Figure 8.12. Substitution of the L tromino that results in a non-periodic tiling.

Figure 8.13. The Ammann–Beenker aperiodic set of two tiles, with substitution rules indicated by heavy black lines, along with a second iteration. Yellow lines enforce matching rules.

Socolar–Taylor aperiodic monotile

The ultimate goal for aperiodic tessellations is the discovery of a single tile that will only tile non-periodically. Such a tile was discovered in 2011 by Joshua Socolar and Joan Taylor [Socolar 2011]. Their tile is not a closed topological disk, however, as a tile was defined earlier in this book. Rather it consists of several disconnect regions, as shown in Figure 8.14. Mirroring is allowed in addition to the rotation of tiles. The discovery of a single aperiodic tile that is a closed topological disk is one of the biggest challenges in tiling theory.

Escheresque tessellations based on aperiodic tiles

M.C. Escher would surely have loved designing lifelike tiles based on the Penrose and other aperiodic tilings. In fact, Escher and Penrose knew and interacted with each other [Schattscheider 2004]. Unfortunately, Escher died in 1972, before the discovery of Penrose tiles and related small sets of aperiodic tiles. Penrose himself created a chicken tessellation from the kite and dart tiles.

From my own efforts at creating Escheresque versions of Penrose tiles, I know that they are particularly difficult templates to work with. A screen print I made in 1993, based on the kites and darts, is shown in

Figure 8.14. A patch of seven Socolar–Taylor aperiodic monotiles. Each of the tiles is given a different color.

Figure 8.15. "Scorpion, Diamondback, and Phoenix", a silk-screen print from 1993.

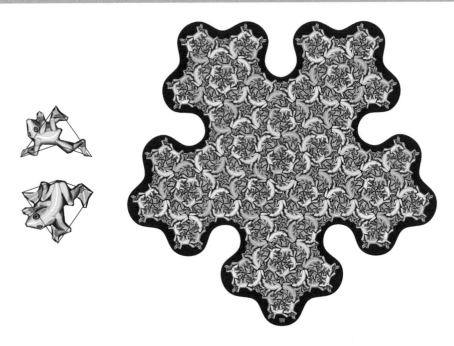

Figure 8.16. "Frog Star II" by Richard Hassell. Copyright 2016 Richard Hassell.

Figure 8.17. Facade of the Sanya Intercontinental Resort, based on the Ammann A2 tiling. Photograph by Patrick Bingham-Hall.

Figure 8.15, where each kite becomes a scorpion and a diamondback rattlesnake and each dart a phoenix bird. Scorpions and rattlesnakes are native animals to the desert near my home in Phoenix, Arizona.

One of the best examples I've seen is the frog tessellation of Richard Hassell (Figure 8.16). Richard is an architect with the Singapore-based firm WOHA, and he incorporated his love for aperiodic tiling in the facade of the Sanya Intercontinental Resort in Sanya, China (Figure 8.17). This is a rare example of an architectural use of Ammann's A2 tiling. Kites and darts are much more common, with numerous examples of floor tiles and pavers around the world.

Activity 8.1. Penrose tiles and the golden number

The golden number appears repeatedly in Penrose tiles. The length of the long edge of the tiles is the golden number times the length of the short edge. In addition, the area of a kite is the golden number times the area of a dart. Finally, in any infinite tiling of kites and darts, the ratio of kites to darts is exactly the golden number.

Materials: Copies of Worksheet 8.1.
Objective: Using Penrose tiles, learn what the golden number is and estimate its value.
Vocabulary: Aperiodic tiles, Penrose tiles, kite, dart, golden number, irrational number.
Activity Sequence
1. Write the vocabulary terms on the board and discuss the meaning of each one.
2. Pass out copies of the worksheet.
3. By measuring the edges of the large kite or dart with a ruler, have the students estimate the value of the golden number.
4. Calculate the ratio of kites to darts in the three patches of tiles on the worksheet. This demonstrates that the larger the tessellation, the better the ratio approximates the golden number.
5. Have the students use calculators to compute the golden number to five decimal places and find the inverse of the golden number and the square of the golden number.

Discussion Questions
1. What value did you get for the ratio of the edge lengths?
2. What values did you get for the ratios of kites to darts for the three patches?
3. What three numbers did you get? How do the three compare? Why do these relationships exist?
4. Using your data for the ratios of tiles in the three patches, make a graph with the number of tiles on the x-axis and ratio on the y-axis. Draw a horizontal line at approximately y = 1.62. What does your graph show?

Worksheet 8.1. Penrose tiles and the golden number

Using a ruler, measure a long edge and a short edge of either the kite or dart. What is the ratio of the lengths?

Count the number of kites and darts in each patch of tiles shown. Calculate the ratio of kites to darts for each patch.

Using a calculator, compute the approximate value of the golden number, $(1 + \sqrt{5})/2$, to five decimal places. Calculate the inverse of the golden number, $1/((1 + \sqrt{5})/2))$. Calculate the square of the golden number.

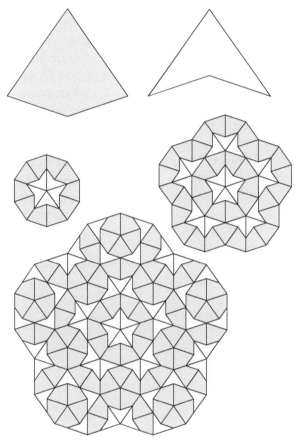

Chapter 9

Fractal Tiles and Fractal Tilings

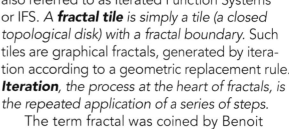

a

b

Unlike "tessellation", there's no concise generally agreed-upon definition for the term "fractal". Broadly defined, *a **fractal** is an object that exhibits self-similarity on progressively smaller scales*. A mathematical fractal has infinitesimally small features. The fractals people are most familiar with are based on equations in the complex plane, most notably the Mandelbrot and Julia sets [Peitgen et al. 1992]. Objects with fractal character abound in nature, for example, mountains, river systems, trees, clouds, coastlines, the arteries in the human body, and distributions such as tree locations in a natural landscape.

There are also many classical fractals that are relatively well known, such as the Hilbert curve, Koch curve and snowflake, and Sierpinski gasket or triangle. These can be referred to as graphical fractals, since they can be readily generated graphically to a reasonable level of detail, without relying on equations or computer coding. These sorts of constructs are also referred to as Iterated Function Systems or IFS. *A **fractal tile** is simply a tile (a closed topological disk) with a fractal boundary.* Such tiles are graphical fractals, generated by iteration according to a geometric replacement rule. ***Iteration**, the process at the heart of fractals, is the repeated application of a series of steps.*

The term fractal was coined by Benoit Mandelbrot in 1975 [Wikipedia 2019], and his book *The Fractal Geometry of Nature* brought the topic to the attention of a wide audience for the first time [Mandelbrot 1982]. Since Escher died in 1972, his knowledge of fractal concepts would have been limited at best.

Since depicting the infinite was a major theme in Escher's work, it seems likely that he would have been fascinated by the infinite level of detail contained in fractals.

Tessellations of fractal tiles

The Koch snowflake or island is an example of a fractal that can serve as a tile. It develops iteratively from a triangle and is the union of three Koch curves. The Koch snowflake tiles with itself using two different sizes, but only in the fractal limit. Figure 9.1 shows how gaps between tiles become progressively smaller as the tile is iterated. In the infinite limit, the area occupied by the gaps goes to zero.

Another example of a tile that tessellates in the fractal limit is shown in Figure 9.2. I initially designed this tile by joining three copies of a hexagonal version of the terdragon fractal curve [Fathauer 2019]. It can also be arrived at by joining tiles in a regular tessellation of hexagons with 1/3 of the tiles missing and then iterating. I've created several other examples of this sort of dendritic fractal tile, using a similar procedure starting from both regular hexagon and square tessellations [Fathauer 2019]. Gaps during the construction steps shrink with each iteration (Figure 9.2), with the untiled area going to zero in the fractal limit. Additional examples of these sorts of dendritic tiles can be seen in Fathauer 2019. Oskar van Deventer used tiles of this sort in puzzle and maze designs [van Deventer 2011].

It's also possible to design fractal tiles that tessellate at each iteration of the construction process. The method used to generate the tile in Figure 9.3 results in a fractal tile that is self-replicating [Vince 1993]. In this specific example, a square is divided into nine smaller squares in such a way as to allow each tile to be replaced by a mirrored copy of a tile with the overall shape of the group, etc. The final fractal tile can be divided into nine copies of itself that are mirrored in addition to being scaled by 1/9 in area. Internal details can be added to such fractal tiles to create Escheresque versions, as shown in Figure 9.4 [Fathauer 2003]. Linear mappings can also be used to generate fractal reptiles [Darst et al. 1998].

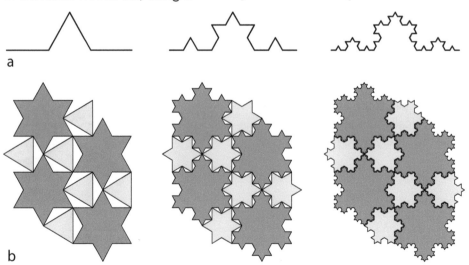

Figure 9.1. (a) Iterative development of a Koch curve. (b) Steps toward tiling of Koch islands shown through three iterations.

Figure 9.2. A dendritic fractal tile generated by joining tiles iteratively in a periodic array. A starting hexagon tessellation minus one-third of the tiles is shown in the upper left, followed by four iterations in forming the tessellation of fractal tiles. A single tile after four iterations is shown in the lower right for clarity.

Figure 9.3. First steps in creating a self-replicating fractal tile by iteratively applying the same steps to a group of nine tiles.

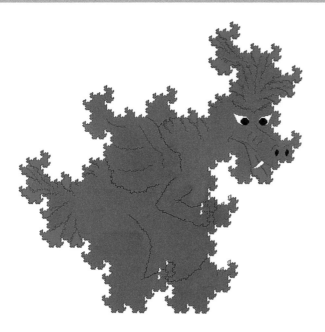

Figure 9.4. An Escheresque version of the tile developed in Figure 9.3.

Fractal tessellations

Another way a fractal character can be incorporated in a tessellation is through the arrangement of tiles. Logarithmic spiral tilings of the sort shown in Chapter 7 are a simple example of fractal tilings, as they possess self-similarity and infinitesimally small features. As noted earlier, Escher executed several tessellation prints that incorporate a singular point at the center, some of which were described in Chapter 7.

In a short treatise published in 1958 entitled "Regelmatige vlakverdeling" ("The Regular Division of the Plane"), Escher shows a portion of a fractal tiling that evolves from a single isosceles right triangle through smaller ones, ending in a line of infinitesimally small triangles, a line of singular points [Ernst 1976]. In the treatise, he adapts a lizard motif to this patch of tiles. He also incorporates this patch of tiles in the only print he made, based on a two-dimensional fractal tiling with more than one singular point, "Square Limit" (1964), as shown in Figure 9.5.

Figure 9.5. Escher's fractal tiling of flying fish. M.C. Escher's "Square Limit" © 2020 The M.C. Escher Company, the Netherlands. All rights reserved. www .mcescher.com.

Note that his "Circle Limit" prints are not fractal tilings, but rather hyperbolic tilings, which are discussed in the next chapter.

Escher turned his earlier patch of isosceles right triangles, shown in Figure 9.6a, into a square bounded by infinitesimally small tiles by using four copies of the patch and filling in corners as shown in Figure 9.6b. Note that in Figure 9.6a, there are locations where two long edges of a tile come together, preventing the placement of additional small triangles against those long edges. A consistent matching rule describing this patch of tiles is to place two half-scale right triangles against the long edge of any right triangle that doesn't meet another long edge. Escher's tiling doesn't follow this rule consistently, but rather adds groups of four triangles, shown in red in Figure 9.6b, to maintain a square boundary. Consistent application of the matching rule

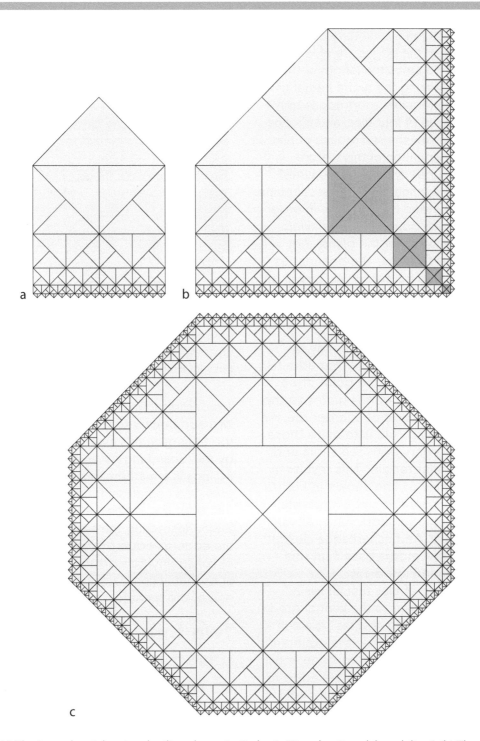

Figure 9.6. (a) The isosceles right triangle tiling shown in Escher's "Regelmatige vlakverdeling". (b) The use of two of the patches from (a) along with corner elements to form a portion of the fractal tiling underlying "Square Limit". (c) The fractal tiling that results from a consistent application of a matching rule.

results in the octagonal boundary of Figure 9.6c, where four of the edges are of length square root of two times the length of the four shorter edges. Making the entire fractal tiling fit in a finite area requires short edges to meet in the center group of four tiles, and it's possible that Escher saw the red groups of tiles as a consistent continuation of this feature.

The "tiling" underlying "Square Limit" doesn't cover the infinite mathematical plane, but rather is bounded in the plane. *In order to distinguish such bounded fractal tilings from tilings that do cover the plane, they will be referred to as* **f-tilings**. The boundaries of f-tilings are often fractal curves, as will be seen in the following examples. f-tilings also differ from "well-behaved" tilings in the fact that they contain singular points.

For a given prototile, there may be more than one matching rule that admits an f-tiling. In Figure 9.6, the point of each tile between two long edges mates to the center of the long edge of a larger tile. An alternative matching rule is to mate these points to the corners between short and long edges of a larger tile. This results in the more complex f-tiling shown in Figure 9.7. This f-tiling contains singular points within the boundary in addition to those on the boundary. An Escheresque design using bat and owl motifs based on this f-tiling is shown in Figure 1.9.

A central group of tiles that allows a bounded f-tiling, similar to the central group of four triangles in "Square Limit", is a common feature of f-tilings. While the example of Figure 9.7 is not strictly edge-to-edge, most of the f-tilings described below are edge-to-edge. In addition, they all contain a single prototile, except for Figure 9.36.

"Square Limit" is the first example of an f-tiling of which I'm aware, not counting simple spiral tilings. Several types of f-tilings

Figure 9.7. An f-tiling using the prototile of Figure 9.6 but with a different matching rule.

are described in the following sections. With two additional exceptions, noted below, all the f-tilings in this chapter were discovered by me. Searching for and discovering new f-tilings is a mathematical recreation that provided countless hours of enjoyable exploration for me, particularly between 1999 and 2006. Restricting the search to single-prototile edge-to-edge f-tilings provided a nice set of constraints to work within. The little discoveries that accompany a search of this sort are one of the great joys of mathematics.

Two-fold f-tilings based on segments of regular polygons

Two infinite families of relatively simple f-tilings can be formed in which the center is a line shared by two congruent tiles that possess a single long edge [Fathauer 2000].

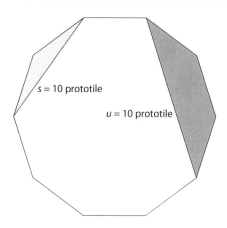

Figure 9.8. The relation of s and u prototiles to regular polygons, in this example, a decagon.

These f-tilings possess two-fold rotational symmetry as well as two lines of mirror symmetry. In the first family, the prototiles are isosceles triangles, while in the second they are trapezoids.

Figure 9.8 illustrates the construction of the prototiles from regular *n*-gons. In the first family, denoted "*s*-type", the tiles are triangular and are generated by connecting two corners encompassing two adjacent edges of a regular *n*-gon. In the second family, denoted "*u*-type", the tiles are trapezoidal and are generated by connecting two corners encompassing three adjacent edges. The f-tilings are constructed by reducing these prototiles in size by a scaling factor given by the ratio of the short to long edges of the prototile and

then matching the long edges of the next-smaller-generation tiles to the short edges of the larger-generation tiles.

When tiles of the same size come together, avoiding overlaps restricts the values of *n* that admit f-tilings of this sort, as shown in Figure 9.9. It turns out *s*-type f-tilings are allowed for all even integer *s* > 2. For *u*-type f-tilings, there is an additional restriction that *u* is of the form 4*m* + 2, where *m* is a positive integer. The first three members of each family of f-tilings are shown in Figure 9.10. The *s* = 4 f-tiling appears to have been discovered by Peter Raedschelders, who used it for an Escheresque tiling of seals [Raedschelders 2003].

Placing two first-generation tiles back-to-back is the simplest starting point or seed, for forming *s*- and *u*-type f-tilings, which can be referred to as "pods". Due to the simple nature of pods, combining them leads to additional attractive f-tilings. In particular, rings and stars can be formed for both *s*- and *u*-type pods. Some examples formed from *s* = 12 pods are shown in Figure 9.11.

Another means of modifying the pods to form new f-tilings is to "perforate" them. This refers to removing the largest tiles and building inward with smaller tiles. This process can be repeated ad infinitum, resulting in complex and delicate filigree-like fractals in some cases (Figure 9.12). Perforation can readily

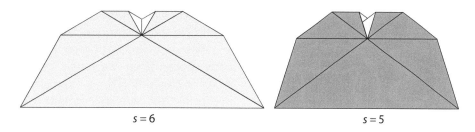

Figure 9.9. Patches of *s*-type f-tilings that are allowed and not allowed by the restriction on tiles overlapping.

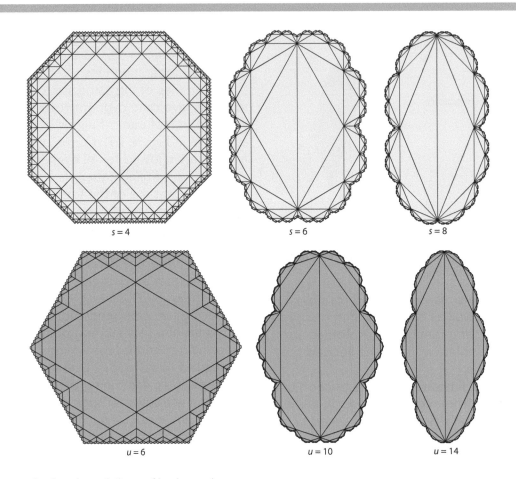

$s = 4$ $s = 6$ $s = 8$

$u = 6$ $u = 10$ $u = 14$

Figure 9.10. The first three *f*-tilings of both *s* and *u* types.

be applied to the other types of *f*-tilings described in the following sections.

The *s* = 6 *f*-tiling lends itself well to Escheresque motifs; some examples are shown in Figure 9.13. Similar motifs could be adapted to larger values of *s*, but the tiles would be increasingly flattened.

Abstract decoration of these *f*-tilings provides another means of creating unique graphic designs. Three examples are shown in Figure 9.14, where invariant mappings were used to adapt designs to the shapes of *s* and *u* prototiles [Ouyang and Fathauer 2014].

Figure 9.11. Some ring- and star-shaped *f*-tilings formed from *s* = 12 pods.

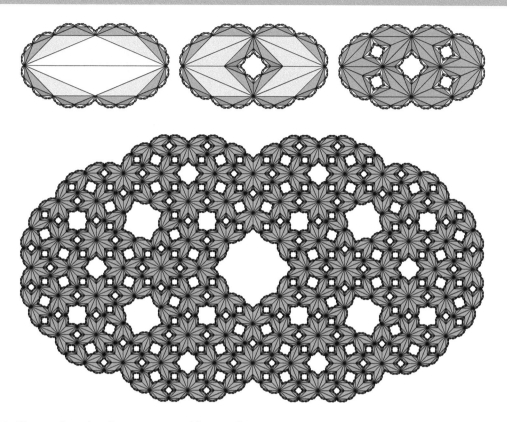

Figure 9.12. The *s* = 8 pod with one, two, and five perforations.

Figure 9.13. Escheresque tiles that allow *s* = 6 *f*-tilings.

Figure 9.14. Three examples of *f*-tilings based on kite-, dart-, and v-shaped prototiles.

f-tilings based on kite-, dart-, and v-shaped prototiles

Edge-to-edge single-prototile *f*-tilings require at least two different edge lengths since a long edge of a scaled-down tile must meet a short edge of a larger tile along its entire length. Kites and darts are simple quadrilaterals that satisfy this requirement. There are families of *f*-tilings for both kites and darts that contain three distinct *f*-tilings each [Fathauer 2001]. V-shaped tiles are closely related to kites and darts, and there are families of *f*-tilings for these containing an infinite number of distinct *f*-tilings [Fathauer 2002].

The three *f*-tilings based on kite prototiles have groups of 6, 8, and 12 tiles at the center as do the dart-based *f*-tilings. The prototiles are shown in Figure 9.15 and the six *f*-tilings in Figures 9.16 and 9.17. The approximate scaling factors for *n* = 6, 8, and 12 are 0.577, 0.414, and 0.268. All other kite and dart *f*-tilings examined have overlaps if the *f*-tilings are built out far enough. The blown-up boundary regions in Figure 9.16 illustrate how the same local structure repeats as additional generations are added, making it clear that overlaps will not occur at any depth of iteration. An Escheresque *f*-tiling of butterflies created by

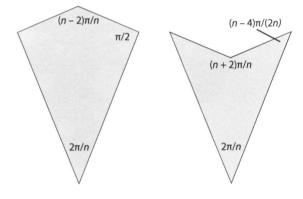

Figure 9.15. Kite and dart prototiles that admit *f*-tilings for *n* = 6, 8, and 12.

Peter Raedschelders in 1985 uses essentially the *n* = 6 kite *f*-tiling [Raedschelders 2003].

The *n* = 8 kite and dart *f*-tilings are two-colorable, but tiles need to be either combined or split to generate two-colorable tilings from the *n* = 6 and 12 *f*-tilings [Fathauer 2012]. Figure 9.18 demonstrates how splitting of the *n* = 6 prototile into two halves results in a two-colorable *f*-tiling that was used for a snake-and-bird Escheresque tessellation. Infinite chains of darts were combined in Figure 9.19 to create a fractal spiral prototile that allows a two-color *f*-tiling.

There are more parameters that can be adjusted in a v-shaped tile than a kite or dart, and this results in a wider variety of allowed *f*-tilings. Only bilaterally symmetric v-tiles

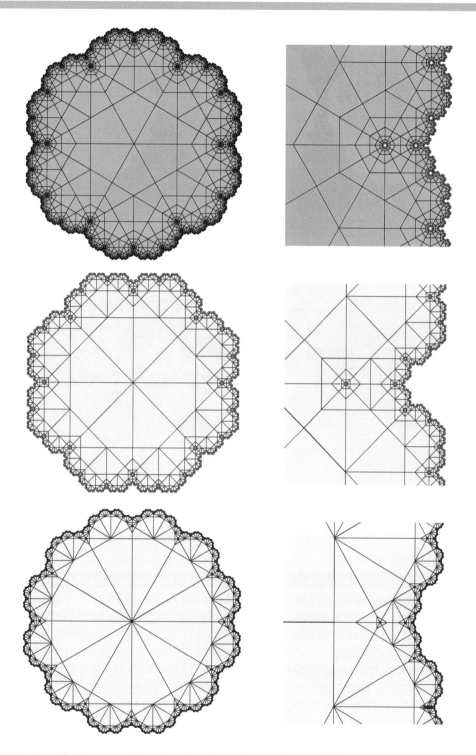

Figure 9.16. The three *f*-tilings possible using kite-shaped prototiles, with blown-up boundary regions illustrating the self-similar structure that allows an infinite number of generations without the overlapping of tiles.

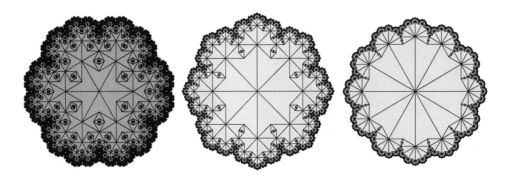

Figure 9.17. The three *f*-tilings possible using dart-shaped prototiles.

Figure 9.18. Three variations on the *n* = 6 kite *f*-tiling. From left to right, the basic *f*-tiling with a different color used for each generation of tiles, a two-colored *f*-tiling made possible by dividing each kite in half, and an Escheresque *f*-tiling of snakes and birds.

(Figure 9.20) will be considered here. The center of an *f*-tiling constructed with such a tile will be a group of *n* first-generation tiles, and the *f*-tiling will have *n*-fold rotational symmetry and *n* lines of mirror symmetry.

There are two ways to fit scaled-down tiles to larger tiles, with the point between two long edges of each smaller tile mated to the *a* points or the *b* points of the large tile. I've only found one edge-to-edge *f*-tiling with

the tips of smaller tiles meeting at *b* points, as shown in Figure 9.21.

There are two infinite families of *f*-tilings with tips meeting at *a* points. For fitting to *a* points, the angle α must be an integer multiple of $2\pi/n$, as must $2\pi - 2\beta$. In the first infinite family, *f*-tilings with *n*-fold rotational symmetry are allowed for all *n* > 2 (Figure 9.22). The boundaries are regular *n*-gons, and the scaling factor between generations is ½.

Figure 9.21. An *f*-tiling in which the tips of smaller tiles mate to *b*-type points in the larger tiles.

Figure 9.19. A digital artwork, "Fractal Tessellation of Spirals" (2011), created by combining tiles in the *n* = 6 dart *f*-tiling to form spiral tiles, allowing a two coloring.

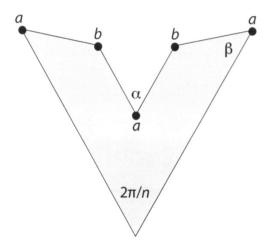

Figure 9.20. A general v-shaped prototile.

In the second infinite family, *f*-tilings with *n*-fold rotational symmetry are allowed for all *n* > 4 (Figure 9.23). The boundaries are fractal curves, and the boundary of the *n* = 6 *f*-tiling is the Koch snowflake. The scaling factor between generations is $1/(4\cos^2(\pi/n))$.

Graphical fractals are in general more interesting when the scaling factor is relatively large, and this holds true for *f*-tilings. The dart-, kite-, and v-shaped prototile *f*-tilings of Figures 9.16, 9.17, and 9.23 all illustrate this point.

There are several other *f*-tilings allowed by v-shaped prototiles. A particularly beautiful example is shown in Figure 9.24. Additional examples can be seen in Fathauer 2002 and Fathauer 2019.

Figure 9.22. The first three members of an infinite family of *f*-tilings based on v-shaped prototiles.

Figure 9.23. The first three members of a second infinite family of *f*-tilings based on v-shaped prototiles.

Figure 9.24. An eight-fold *f*-tiling based on a v-shaped prototile that is not a member of the families of Figures 9.22 and 9.23, with a blowup of one of the boundary inlets.

f-tilings based on polyforms

The three regular tessellations are based on the square, the equilateral triangle, and the regular hexagon. In order to find new tessellations such as f-tilings, these polygons would seem to be a promising starting point. It turns out that polyiamonds, polyominoes, and polyhexes are a rich source for uncovering edge-to-edge, single-prototile f-tilings. In particular, dissecting these polyforms allows a systematic search that generates a wide variety of f-tilings [Fathauer 2009]. Some examples are given here, but a wider selection can be found in Fathauer 2009 and Fathauer 2019.

An infinite family of f-tilings can be formed from polyominoes consisting of n unit squares arranged in a straight line, as shown in Figure 9.25. Note that the single square yields the same f-tiling as $s = 4$ above. The long edges of the first-generation tiles are given by the square root of $1 + n^2$ and the small angles by the arctangent of $1/n$. These values are for the most part irrational, but the tiles fit when scaled and rotated by the right amounts.

More complex polyominoes offer more prototile candidates. For example, dissection of the plus pentomino reveals three possibilities for distinct prototiles, but only one of the three admits an f-tiling (Figure 9.26).

More interesting f-tilings are possible using larger polyominoes, in some cases forming nice enclosed regions. In Figure 9.27, dissection of a 16-omino results in the four-fold f-tiling shown. The tiles fold together to form enclosed plus-pentomino holes, with the number of holes quintupling with each iteration. The holes fill in the infinite limit.

Similar to the polyomino case, there is an infinite family of f-tilings allowed by linear chains of 1, 2, 3, 4, … hexagons, the first of which is the $u = 6$ f-tiling (Figure 9.10). If three hexagons are joined not in a line but with three-fold symmetry, two prototiles are possible (Figure 9.28a). One is the v-shaped tile of Figures 9.21 and 9.22. The other can be mated to larger tiles at two distinct points, as shown in Figure 9.28b. In addition, a prototile lacking mirror symmetry can be mirrored between successive generations to create distinct f-tilings. For the prototile of Figure 9.28, there are thus four possible matching rules, leading to four distinct f-tilings. Three of these are shown in Figure 9.29, and they are seen to have very different boundaries. The fourth possibility has a similar boundary to the bottom right f-tiling in Figure 9.29.

The left f-tiling in Figure 9.29 has an unusual and intriguing wavelike boundary. I took advantage of this feature in a digital artwork, in which the individual tiles are

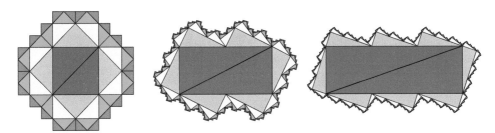

Figure 9.25. The first three members of an infinite family of f-tilings derived by bisecting bar polyominoes carried through five generations of tiles.

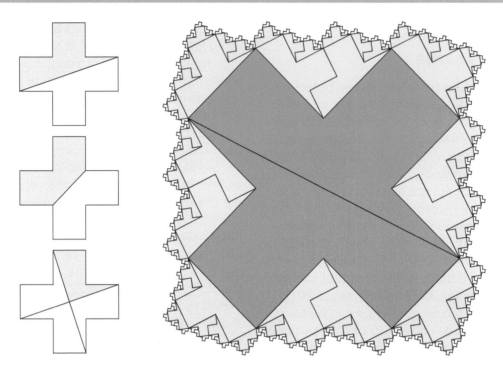

Figure 9.26. The three candidate prototiles and one allowed *f*-tiling based on dissecting the plus pentomino.

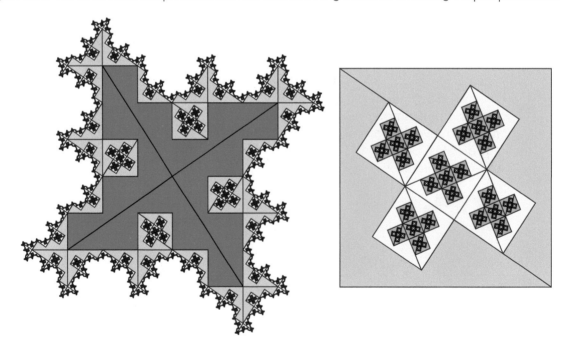

Figure 9.27. An *f*-tiling based on a dissection of a 16-omino, with a blown-up patch showing the formation of plus-pentomino-shaped holes that fill in the limit.

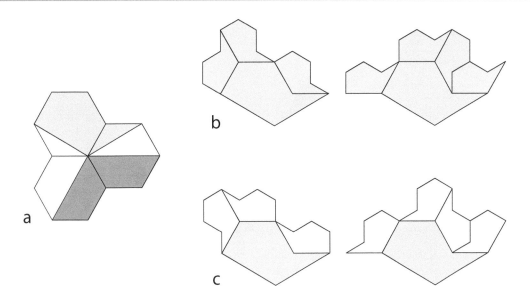

Figure 9.28. (a) The two prototiles that can be created by dissecting a three-fold 3-hex. (b) Smaller tiles mating to a large tile at two distinct points. (c) Mirrored smaller tiles mating to a large tile at two distinct points.

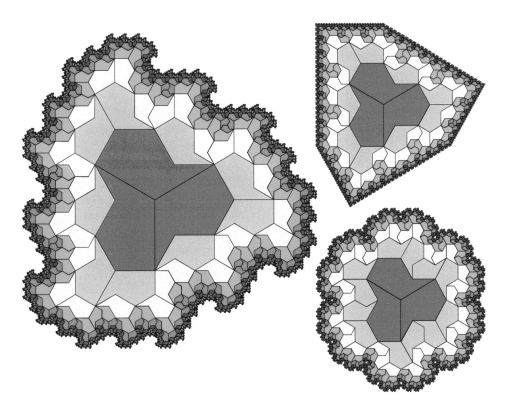

Figure 9.29. Three *f*-tilings allowed by the prototile and matching rules of Figure 9.27.

Figure 9.30. "Fractal Fish – Grouped Groupers" (2001), a tessellation artwork based on one of the *f*-tilings of Figure 9.29.

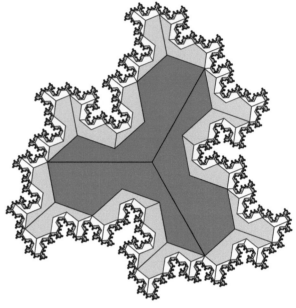

Figure 9.31. An *f*-tiling based on a dissection of a 4-hex.

groupers and the tiling of groupers can also be interpreted as waves (Figure 9.30).

Two more examples of *f*-tilings based on dissections of polyhexes are shown in Figures 9.31 and 9.32. For both prototiles, there are four possible matching rules, only one of which is shown. In Figure 9.31, both the prototile and boundary of the *f*-tiling have three-fold rotational symmetry but lack mirror symmetry. In Figure 9.32, the boundary has lines of mirror symmetry even though the prototile does not. This *f*-tiling, based on a 9-hex, has a particularly beautiful inlet in the boundary.

Polyhexes are really a subset of polyia-monds, since a hexagon can be constructed from six equilateral triangles. There are infi-nitely many polyiamond-based *f*-tilings apart from those that are also polyhexes, though.

These include the two-fold $s = 6$ *f*-tiling of Figure 9.10 and the six-fold *f*-tiling of Figure 9.17. There are also three-fold polyiamond-based *f*-tilings, but the most interesting examples have six-fold symmetry.

The prototile in the *f*-tiling of Figure 9.33 is created by dissecting a 36-iamond. Some *f*-tilings, this example included, form a super-tile which tessellates by translation. In Figure 9.33, portions of two adjacent supertiles are shown. The channels where the three come together will close up in the infinite limit.

The prototile in the *f*-tiling of Figure 9.34 is created by dissecting a 24-iamond. This is an example of a pseudo-edge-to-edge *f*-tiling. If the point A is considered a corner of the large light-blue tile, then the *f*-tiling can be considered edge-to-edge. This *f*-tiling has particularly intricate and beautiful inlets, as shown in more detail in the blown-up region.

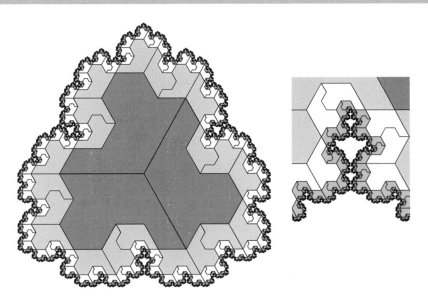

Figure 9.32. An *f*-tiling based on a dissection of a 9-hex, with a close-up view of one of the inlets.

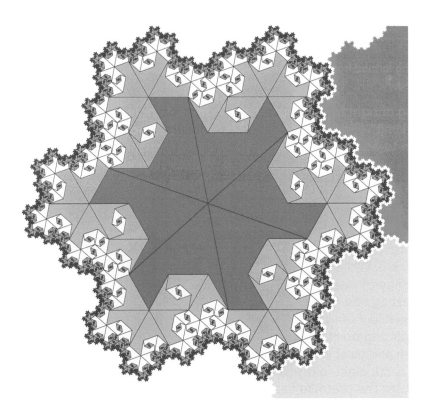

Figure 9.33. An *f*-tiling formed from a dissected 36-iamond. The light red and blue tiles are additional copies of the *f*-tiling that tessellate with the full *f*-tiling.

Figure 9.34. An *f*-tiling formed from a dissected 24-iamond. The blown-up region highlights the intricate nature of the inlets in the boundary.

Note that the tiling of right triangles used in Escher's "Square Limit" (Figures 9.5 and 9.6) is also pseudo-edge-to-edge.

Miscellaneous *f*-tilings

There are innumerable *f*-tilings not included in the above categories. Polyominoes, polyhexes, and polyiamonds have angles that are multiples of $\pi/2$ or $\pi/3$. While these angles offer the widest range of possibilities, other angles yield interesting *f*-tilings as well. An example is the two-fold *f*-tiling shown in Figure 9.35. The starting point for this prototile is a bowtie with angles of $\pi/5$ and $3\pi/5$. Another example is the eight-fold *f*-tiling shown in Figure 9.36. The prototile has 14 edges, and the angles are all multiples of $\pi/4$.

Without restricting the discussion to *f*-tilings that are edge-to-edge and constructed from a single prototile, the number of possibilities becomes unmanageable. However, constructions based on very simple prototiles such as squares and equilateral triangles are worth considering. These can't be

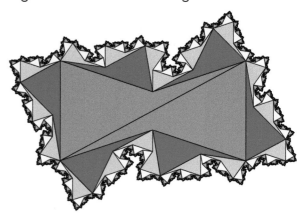

Figure 9.35. A two-fold *f*-tiling based on a dissection of a bowtie with angles of $\pi/5$ and $3\pi/5$.

Figure 9.36. An eight-fold *f*-tiling with prototile angles that are multiples of π/4.

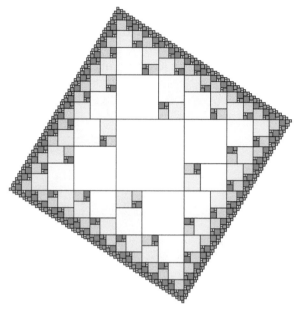

Figure 9.37. A fractal tiling of squares based on the golden spiral.

edge-to-edge since all the edges are of the same length. Some pseudo-edge-to-edge examples can be constructed by mating two or three smaller tiles to a large tile, though these tend to not be particularly attractive. A more subtle construction is shown in Figure 9.37, where golden spirals appear an infinite number of times. Starting from a central square, golden-ratio-scaled squares are iteratively arranged until they become, infinitesimally small at the boundary as well as singular points inside the boundary. Eight generations of squares are shown in the figure.

When two prototiles are allowed, the possibilities again become unmanageably large. Two prototiles that are both simple and highly symmetric polygons yield some *f*-tilings that are noteworthy, such as that shown in Figure 9.38, though consisting of hexagrams and isosceles triangles. Phil Webster has used *f*-tilings of this sort as the basis for fractal Islamic tilings [Webster

Figure 9.38. An *f*-tiling of hexagrams and isosceles triangles.

2013]. Peichang Ouyang has used hierarchical subdivision of adjacent tiles to generate a wider variety of *f*-tilings than can be constructed using simple matching rules [Ouyang et al. 2018].

Finally, it's possible for both the prototile and the tessellation to be fractal. An example is an *f*-tiling of Koch snowflakes, shown in Figure 9.39.

Figure 9.39. An *f*-tiling of Koch snowflakes.

Activity 9.1. Prototiles for fractal tilings

Materials: Copies of Worksheet 9.1.

Objective: Understand how modifying polygons can lead to prototiles that tessellate in a fractal fashion.

Vocabulary: Fractal tiling, prototile.

Activity Sequence

1. Write the vocabulary terms on the board and discuss the meaning of each one.
2. Pass out copies of the worksheet. Show Figure 9.29 on a projector.
3. Have the students work through the questions on the worksheet.
4. Have the students share their results on how they arrived at their answers.

Discussion Questions

1. What result did you get for the area and why?
2. What values did you get for the angles α and β? How did you arrive at your answers?
3. Share your value for α computed using the two different approaches and how you arrived at that value.
4. Who managed to calculate the infinite series? Share your reasoning and steps.

Worksheet 9.1. Prototiles for fractal tilings

The dashed lines in the leftmost figure show how the prototile of the fractal tilings in Figure 9.28 is related to a regular hexagon. How does the area of the tile compare to the area of the hexagon?

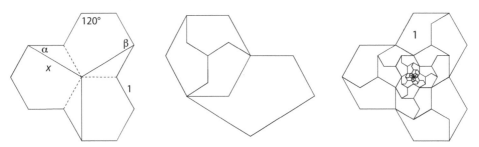

If the tile's short edges and the hexagon's edges have length 1, how does the area of the tile compare to the area of an equilateral triangle with edge length 1?

Given that the interior angles of a hexagon are each 120°, what are the values of the angles α and β?

Given that the sine of α is 0.5, use the Pythagorean theorem to deduce the value of x, the length of a long edge of the prototile.

Note that x is also the linear scaling factor between two successive generations of tiles. The area between two successive generations of tiles will scale as x^2. Using this fact, determine x by comparing the middle and leftmost figures.

For the collection of tiles shown in the rightmost figure, given the area of one of the large tiles is 1, what is the total area of the infinite number of tiles shown?

Use your answer to the last question to calculate the sum of the infinite series $1/3 + 1/3^2 + 1/3^3 + 1/3^4 + \ldots$

Chapter 10

Non-Euclidean Tessellations

Number of polygon sides n →

{6, 3} | {6, 4}

dodeca-
hedron
{5, 3} | {5, 4} | {5, 5} | {5,

cube
{4, 3} | {4, 4} | {4, 5}

tetra-
hedron
{3, 3} | octa-
hedron
{3, 4} | icosa-
hedron
{3, 5}

Number of polygons k meeting at a vertex →

A ll the geometric tessellations described in the preceding chapters have lived in the Euclidean plane. This is the geometry that most people use in daily life without it being explicitly stated. Elementary geometry problems, blueprints used by architects and construction workers, and physical and computer desktops all deal with flat surfaces. In Euclidean geometry, parallel lines never meet, and the sum of the angles in tiles meeting at a point is 360°. Euclidean geometry is named after the ancient Greek mathematician Euclid, who laid out the axioms on which it's based.

However, there are other geometries, with their own axioms, and it's enlightening and entertaining to consider tessellations in them. *Geometries in which the angles about a point sum up to less than 360° are called* **elliptic**, *while geometries in which the angles about a point sum up to greater than 360° are called* **hyperbolic**. *The most commonly considered elliptic geometry is* **spherical geometry**, *which takes place on the surface of a sphere.* This may seem abstract, but we live on a sphere. We normally think in terms of Euclidean geometry simply because the Earth is so large compared to us that it appears to be a plane in our everyday experience. Treating it as such is a very good approximation for most purposes.

Even on a local scale our world isn't Euclidean. If it were, the ground would be perfectly flat. The hills, valleys, and undulations of a surface can be described by the curvature at each point. Curvature can be positive, zero, or negative. In Euclidean geometry, where surfaces are perfectly flat everywhere, the curvature is zero. *A surface has* **positive curvature** *at a point if the surface curves away from a*

(3.5.3.5)

tangent plane in the same direction in any cutting plane, while it has **negative curvature** if it curves away in different directions. An example of positive curvature is a point on a sphere, and an example of negative curvature is a point on a saddle or the bell of a trumpet.

Regular tessellations can be understood in a more comprehensive fashion if all three geometries are considered together. Recall that a regular tessellation is an edge-to-edge tessellation by regular polygons. *A regular tessellation can be specified by the integer n for the number of sides of each polygon and k for the number of polygons meeting at each vertex via the* **Schläfli symbol** *{n, k}.* The regular plane tessellation of hexagons, e.g., is specified by {6, 3}. This symbol can also be applied to elliptic and hyperbolic tessellations. All possibilities for n and k between 3 and 6 are listed in Figure 10.1. Note that the product $(n − 2)(k − 2) = 4$ for the plane tessellations, while it's greater than 4 for hyperbolic and less than 4 for elliptic tessellations [Coxeter 1954].

For the regular tessellations in the plane, recall that the {3, 6} tessellation is dual to the {6, 3} tessellation, and the {4, 4} tessellation is self-dual. This pattern holds for elliptic and hyperbolic tessellations as well. For any {p, q}, the dual tessellation is {q, p}, with the self-dual case corresponding to p = q. For polyhedra, the cube is dual to the octahedron, the dodecahedron to the icosahedron, and the tetrahedron to itself.

Hyperbolic tessellations

In hyperbolic geometry, the curvature of a surface is negative everywhere. This means every point on a hyperbolic surface is a saddle point. Models can be constructed of hyperbolic surfaces embedded in Euclidean three-space. One way to do this is by crocheting outward from a starting point such that the increase in the number of stitches is exponential as one moves outward [Taimina 2018]. Exponential growth can also be seen in nature in sea life like flatworms and some corals, in plants like kale and green-leaf lettuce, and in some fungi. As these organisms grow, there is excess material that can't lay flat, leading to rippled surfaces.

Similarly, a hyperbolic tiling can be constructed in our three-dimensional world, and the same sort of rippling is observed. As an example, an f-tiling of dart-shaped tiles is shown in Figure 10.2, in which the angles of the tiles meeting at each vertex exceed 360° [Fathauer 2014]. The wavy features of this f-tiling are similar to the features observed in hyperbolic crochet and natural exponential growth surfaces. Note that the concept of curvature only applies to smooth curves, not constructs of flat tiles.

An infinite hyperbolic tessellation can be depicted in the plane using a suitable model. The most common method is the Poincaré disk or conformal disk model used in Figure 10.1. In this model, everything is contained in a unit disk. Straight lines appear as either segments of circles that are perpendicular to the disk at its boundary or diameters of the disk. Straight edges of tiles appear curved in this model, with tiles being larger near the center and smaller near the edges.

Tilings that are not possible in the Euclidean plane are allowed in the hyperbolic plane. For example, regular heptagons can tile the hyperbolic plane. A tiling could also be constructed with equilateral triangles such that seven or more triangles meet at each vertex. Three two-colorable tilings are shown in Figure 10.3, with alternating regular polygons of two types [Wikipedia 2019]. The same tessellation can look different depending on where the image is centered in the disk. For

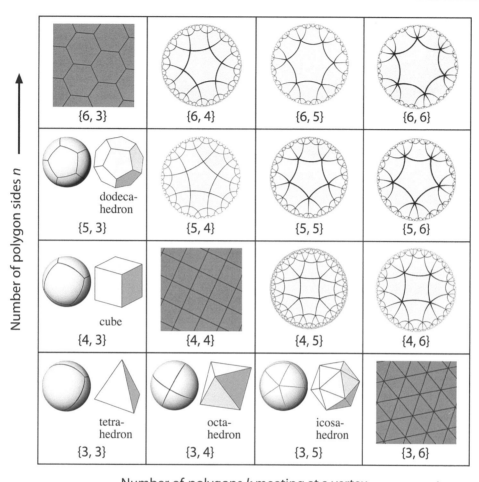

Number of polygon sides *n*

Number of polygons *k* meeting at a vertex

Figure 10.1. Regular tessellations for the number of polygon sides *n* = 3 to 6 and number of tiles meeting at each vertex *k* = 3 to 6, labeled with the Schläfli symbol {*n*, *k*}. Elliptic tessellations are yellow, Euclidean red, and hyperbolic blue. For larger *n* and *k*, the tessellations are all hyperbolic.

example, three representations of the {3, 7} tiling are shown in Figure 10.4.

A common theme in M.C. Escher's art was approaches to infinity [Ernst 1976]. Escher discovered the Poincaré disk in a book by H.S.M. Coxeter [Ernst 1976]. It allowed him to depict an infinite tessellation on a finite piece of paper, as seen in his four "Circle Limit" prints. In "Circle Limit III" (1959), the flying fish travel along curved lines from one point on the boundary to another, with fish becoming infinitesimally small at either end. The geometry of this print is similar to that of the (3.4.3.4.3.4) tiling, though it's not an accurate representation of the tiling. Doug Dunham has created hyperbolic tessellations of Escheresque flying fish with other symmetries [Dunham 2006]. Jos Leys has created hyperbolic tessellations from many of Escher's plane tessellations [Leys 2019].

The upper-half-plane model and the band model are two other popular representations

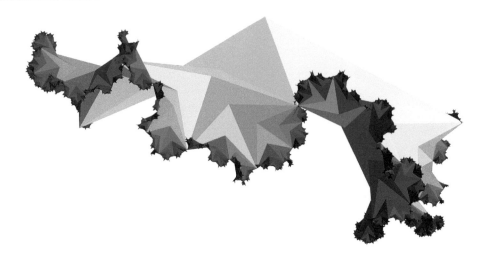

Figure 10.2. A hyperbolic *f*-tiling of dart-shaped tiles exhibits the wavy features characteristic of hyperbolic surfaces embedded in Euclidean three-space.

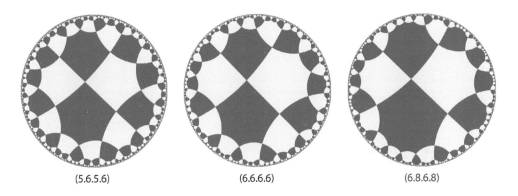

(5.6.5.6) (6.6.6.6) (6.8.6.8)

Figure 10.3. Hyperbolic tilings with four regular polygons meeting at each vertex.

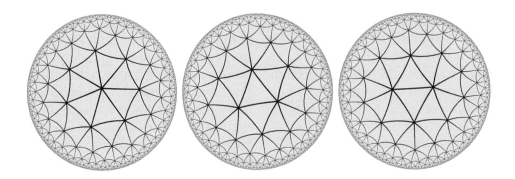

Figure 10.4. The {3, 7} tessellation depicted in the Poincaré disk model, centered on three different points.

of the hyperbolic plane. In addition, squares and other boundaries are possible. These and other conformal models are described in detail in a talk by Vladimir Bulatov that includes several nice animations [Bulatov 2010]. In conformal models, the mapping between Euclidean and hyperbolic planes preserves orientation and angles locally. Some models of the hyperbolic plane, such as the Klein disk model, are not conformal.

Hyperbolic tilings can be quite beautiful. They have been raised to an art form by Roice Nelson in his TilingBot tweets, many of which are animated [Nelson 2019]. Three examples are shown in Figure 10.5. In these images, the edges are sometimes widened and blurred and vertices are sometimes given a filled circle or glow. Modifications such as truncation and rectification are sometimes shown as well. The smaller triangles are primitive unit cells for the tiles, with the two mirror orientations typically given different colors.

Models of the hyperbolic plane are transformations that change the appearance of a tiling, and different transformations can be combined to yield new images. The result of

a

b

c

Figure 10.5. Renderings of three hyperbolic tilings of regular polygons. (a) The {3, 10} tiling is shown in the upper half-plane model. (b) The {8, 3} tiling is shown in green and white, and the second truncation shown in diffuse black lines using the conformal square model. (c) The {8, 3} tiling is shown in black and gray and its rectification in white using the band model. Images by the twitter bot @TilingBot.

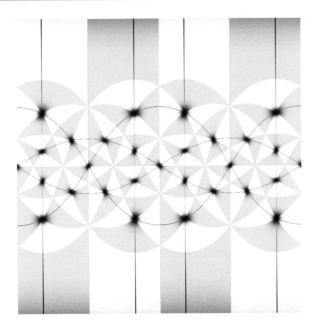

Figure 10.6. Mercator projection of the {3, 5} tiling, along with black lines showing the rectification of the tiling. Image by the twitter bot @TilingBot.

such operations may not be a model of the hyperbolic plane but can still be interesting from a mathematical or an aesthetic standpoint. The starting point for such transformations doesn't have to be a Euclidean tiling. An example is shown in Figure 10.6, where the spherical tiling {3, 5} was the starting point [Nelson 2019]. The interested reader can explore further by looking into Möbius transformations and projective geometry.

Spherical tessellations

A **spherical tessellation** *is a tessellation on the surface of a sphere; i.e., a collection of shapes that cover the surface of a sphere without gaps or overlaps.* Spherical tilings are closely related to polyhedra, and convex polyhedra can be projected onto the surface of a sphere. In Figure 10.1, both the polyhedral and spherical versions of the elliptic regular tessellations

are shown. The classic black-and-white soccer ball is essentially a spherical truncated icosahedron. Three other spherical Archimedean solids are shown in Figure 10.7 [Wikipedia 2019].

Spherical tessellations are another way of avoiding boundaries in the depiction of a tessellation. Escher carved a few wooden spheres with adaptations of his plane tessellations, including fish and reptile motifs, as well as angels and devils [Schattschneider 2004].

It's straightforward to construct tessellated polyhedra from paper, as described in Chapters 21–24. There's no simple way to construct tessellated spheres, however, since the surfaces are curved. Two-dimensional representations of spherical tessellations can be coded using software such as KaleidoTile [Weeks 2019], which I used to produce the spherical tessellations of Figure 10.8. Spherical, planar, and hyperbolic tessellations of sea stars and scallops using the same primitive unit cell are shown in Figure 10.9.

As discussed in Chapter 3, Frieze groups classify symmetries in one dimension, while wallpaper groups classify symmetries in two dimensions. In three dimensions, spherical symmetry groups, also known as point groups, are used. Since this book is mainly concerned with tessellations in the plane, the spherical groups will not be described in detail. There are several systems for classifying these, one of which is the orbifold notation described in Chapter 3 for two dimensions. Note that rotation is about an axis in three dimensions, in contrast to a point in two dimensions. Similarly, reflection is about a plane in three dimensions, as opposed to a line in two dimensions.

As an example, consider the regular icosahedron. There is one distinct point where five mirror planes meet and through which a five-fold axis passes, one distinct point where three mirror planes meet and through

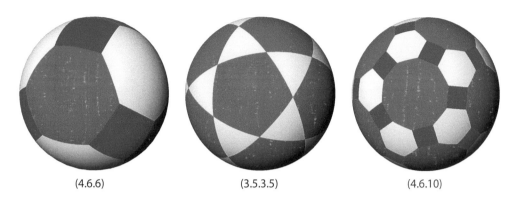

(4.6.6)   (3.5.3.5)   (4.6.10)

Figure 10.7. Three spherical Archimedean tilings.

Figure 10.8. Three different spherical tessellations of flowers and leaves created from the same triangular primitive unit cell using KaleidoTile.

Figure 10.9. Spherical, planar, and hyperbolic tessellations of sea stars and scallops created from the same triangular primitive unit cell using KaleidoTile.

Figure 10.10. A regular icosahedron without surface decoration, with a surface tessellation, and projected onto a sphere to form a spherical tessellation. The spherical symmetry group is the same for all three. Images made with KaleidoTile.

which a three-fold axis passes, and one distinct point where two mirror planes meet and through which a two-fold axis passes (Figure 10.10). The orbifold signature is *532, where the asterisk indicates reflection. If the icosahedron is decorated with tessellation graphics that preserve the mirror symmetry, such as the sea star and scallop tessellation of Figure 10.9, the symmetry group is the unchanged. If it's projected onto the surface of a sphere, the symmetry group remains unchanged.

Activity 10.1. Non-Euclidean tessellations of regular polygons

Materials: Copies of the worksheet.

Objective: Learn how to determine if a regular tessellation is elliptic or hyperbolic, how to specify them using Schläfli symbols, and how to describe and quantify vertex types in hyperbolic tessellations.

Vocabulary: Elliptic tessellation, hyperbolic tessellation, Schläfli symbol, two-colorable.

1. Write the vocabulary terms on the board and discuss their meanings.
2. Pass out copies of the worksheet.
3. Have the students complete the worksheet. It might be a good idea to check their responses to the first question before having them complete the other two.

Discussion Questions

1. Can you think of any real-world examples of elliptic tessellations?
2. Which Platonic solid is the third tessellation of Figure 10.11 the spherical version of?
3. Why are some of the tessellations in Figure 10.12 two-colorable, while some are not?

Worksheet 10.1. Non-Euclidean tessellations of regular polygons

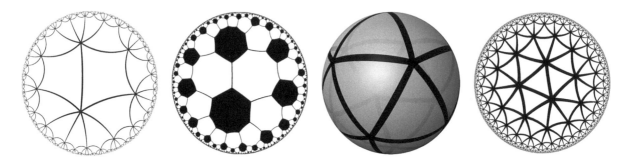

In the Euclidean plane, there are three regular tessellations: {3, 6}, {6, 3}, and {4, 4}. Are the following regular tessellations elliptic or hyperbolic?

Five squares meeting at each vertex:

Three octagons meeting at each vertex:

Five triangles meeting at each vertex:

Three pentagons meeting at each vertex:

Write the Schläfli symbol under each of the regular tessellations shown in Figure 10.11.

Write the vertex description under each hyperbolic tessellation of regular polygons shown in Figure 10.12, and also calculate the angle sum at each vertex.

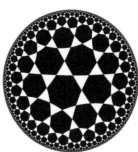

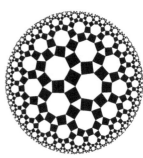

Chapter 11

Tips on Designing and Drawing Escheresque Tessellations

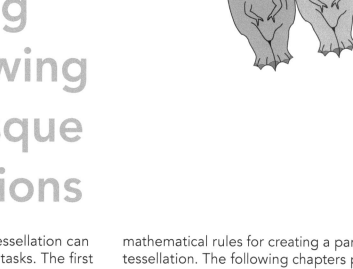

Creating an Escheresque tessellation can be divided into two main tasks. The first is designing the prototile or prototiles, the tile or tiles to which all of the tiles in the tessellation are similar. The second is making multiple copies of the prototile(s) to form the full tessellation. The second task is relatively straightforward once you understand the mathematical rules for creating a particular tessellation. The following chapters provide numerous templates, as well as grids, to help you create tessellations. These templates describe exactly how the rules work for a wide variety of tessellations. A detailed example is described in Chapter 13, walking you through each step in the process.

The first task, designing your prototile(s), is probably the more challenging part of making a good tessellation. It involves a lot of creativity, as well as patience and perseverance. It's something that most people find very difficult at first, but with practice it becomes much easier. The task of designing your prototile(s) has two facets, designing the outline of the tile(s) and drawing the internal details. These two tasks go hand-in-hand when creating a prototile. You will probably find yourself going back and forth between the two as you refine your design.

The first step in designing a prototile is to distort the edges of the tile. You need to exercise your creativity to recognize a motif emerging from a tile that you're distorting. You then guide further distortions to make the tile look more like that motif. In a tessellation, every line segment is shared by two tiles. This means that changing a line to make a tile look better in one area could make another portion of the tile (or of a second tile) look worse.

Designing an Escheresque tile is easier if you choose motifs that are flexible. Some animals can be twisted around and distorted quite a bit, while others can't. Lizards, birds, and fish seem to work particularly well, and Escher used these motifs for many of his tessellations. As a motif, people are especially challenging. Not only are they not very flexible, but the human eye is more sensitive to whether or not people look right than to whether or not other animals that we don't see as often look right.

The purpose of this chapter and the next is to provide guidelines and tips to help you create the best possible prototile(s) and tessellations. The first four tips in this chapter layout goals to strive for in order to realize an effective tessellation. Achieving all these design goals in the same tessellation is often not possible, and you'll notice that many of the tessellations in this book fall short of these ideals. Tips 5–8 provide ways to improve your

designs. Chapter 12 describes special tricks and techniques to help you solve problems or further improve your designs.

Drawing tessellations by hand

Some people create tessellations completely by hand, some completely by computer, and some using a mixture of the two. How you choose to work depends on your background, skillset, and personal preferences. Different approaches are compared and contrasted below.

It is entirely possible to create unique and beautiful tessellations without the use of computers. M.C. Escher worked this way, and some people still prefer to do so today. However, Escher was a graphic artist by training, and computer graphics were virtually non-existent until late in his life. On the positive side, working by hand frees you up to draw in a more relaxed and spontaneous fashion. This may help you to be more creative. A hand-drawn and colored tessellation may look more artistic than the one created by a computer, as these hand techniques are the traditional means of creating art. On the negative side, it is difficult to accurately draw the same lines over and over again, particularly when they are reflected. It can also be tedious to repeatedly draw the same features in order to tessellate a large area. If your drawing skills are not very developed, it takes even longer and can be more difficult to achieve satisfactory results working by hand.

Using general computer graphics programs

Tiles can be created using drawing programs that are not specifically designed for tessellations. These programs allow easy duplicating, aligning, rotating, translating, and reflecting of lines and other elements. All the templates in the later chapters can easily be set up using such a program. While it is more difficult to draw interior details using a computer, they only need to be drawn once, after which the entire tile with interior details can be quickly duplicated and transformed to create a larger tessellation. This is how I personally create tessellations, using the program FreeHand. Unfortunately FreeHand is no longer being maintained and won't run with newer operating systems. A similar program that is widely used by graphics professionals is Illustrator, and CorelDRAW is another popular drawing program. After I have created a tessellation using FreeHand, I will copy the tiles and paste them into Photoshop to perform additional rendering for final artwork. FreeHand and Illustrator are vector-based programs, while Photoshop is pixel based. There are many drawing programs for smartphones and tablets as well, but I personally prefer the control afforded by a computer.

Using a tessellations computer program

Computer programs have been written specifically for creating tessellations. An early program that has been used in many classrooms is *TesselMania!* by Kevin Lee. An updated version that runs on modern machines and operating systems is called TesselManiac! (note the "c") [Lee 2016]. There are also some online tools for creating Escheresque tessellations. It's easier to create a tessellating tile using these programs, which automate the duplicating, rotating, and other operations. They are also helpful in visualizing and understanding transformations. They provide easy coloring of tiles and also drag-and-drop interior

features like eyes. I personally find these programs limiting, however, so I have not used them for creating my own tessellations. They have a limited number of templates and limited control over the shaping of lines, shading, etc., compared to programs like Illustrator and Photoshop. They don't allow the same degree of polish and control over the appearance of the final tessellation as professional graphics programs.

Mixing techniques

The above approaches can be combined, of course. You might want to use a tessellation program to design your tile shape and then create a final version using other software. Or you might design your tile shape using a drawing program and then print the outlines of the shapes and draw the interior details by hand. You might create both the tile shape and interior details using a computer and then just do the coloring by hand. You should try different approaches and decide what works best for you. Regardless of how you design your prototiles, the following tips can help improve your designs.

Tip 1: The outline of the tile should suggest the motif

It's often possible to make a shape that doesn't particularly look like a lifelike motif and make it take on life with the interior details. While this sort of tessellation can be effective, it's more satisfying if the shape by itself conjures up a motif. In the examples of Figure 11.1, most people wouldn't think *Tyrannosaurus rex* if they saw the left shape by itself. On the other hand, a winged dragon or something similar is immediately

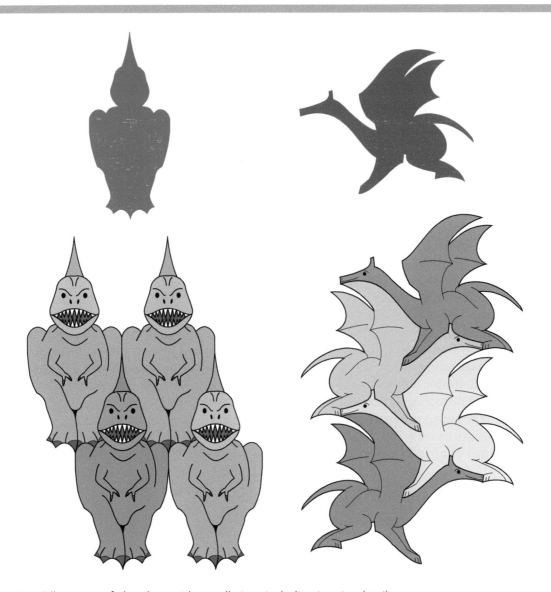

Figure 11.1. Silhouettes of tiles along with tessellations including interior details.

suggested by the right shape. It usually takes more work to create a shape like that on the right, and the resulting tessellation is more compelling because the individual tiles are more interesting. You might want to look at some of Escher's tessellations to see which ones have outlines that suggest the motif and which ones need the interior details to guide your interpretation of the tile(s).

Tip 2: The tiles should make orientational sense

In a tessellation, like most artwork, you are creating a scene and to a degree telling a story. That scene will look more natural if the motifs are oriented in a way that is consistent with the

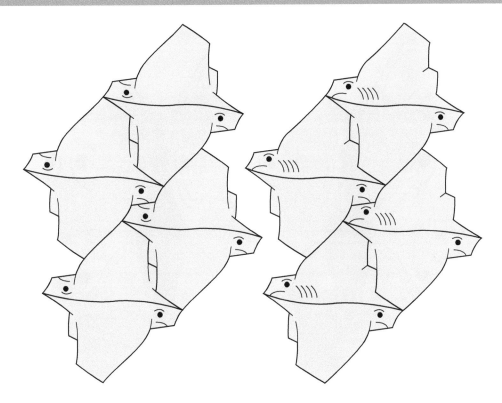

Figure 11.2. Tessellations of rays contrasting designs, in which half the rays are upside down and in which all the rays are right-side up.

world around us. Most of the time, your motifs will either be drawn from a top view or a side view. Whichever it is, the two shouldn't be mixed, unless you're trying to achieve a special effect. If you're using a side view, then you usually will want the ground at the bottom and the sky at the top, even if these are only implied, not actually shown. Another way of stating this is to say that the pull of gravity should be oriented the same for all of the tiles.

The example of Figure 11.2 shows two tessellations of rays. In the left version, each tile is identical. As a result, half the rays are swimming right-side up and half upside down. In the right version, the shapes of half the tiles have been modified slightly and the interior details changed so that all of the rays are swimming right-side up. This makes for a more effective design.

As you design a tessellation with a side-view perspective, you may need to reorient your entire tessellation, rotating the paper or rotating all of the tiles on your computer screen, in order to get the pull of gravity in the downward direction. If you look at Escher's tessellations, you'll find that almost all of them make orientational sense, other than his very earliest attempts.

Tip 3: Choose motifs that go together

A tessellation makes more sense, in effect tells a better story, if all the motifs go together. They can be either similar or contrasting. The example of Figure 11.3 has three motifs – beetles, moths, and bumblebees, all

Figure 11.3. A tessellation of three different types of insects.

three of which are insects. Note that they are all seen from a top view, so the tessellation also makes orientational sense. An example of contrasting motifs is angels and devils, used by M.C. Escher in one of his most popular prints. Another example by Escher is the use of birds and fish, contrasting the elements sky and water. It wouldn't make sense to use motifs that have no clear relationship to each other, such as dinosaurs along with flowers, or boats along with cats. When Escher used boats, he used them with fish. The Escheresque tessellation puzzles I designed combine like motifs, such as *Squids and Rays*; *Beetles, Moths, and Bumblebees*; *Frogs, Lizards, and Snakes*; and *In the Garden*.

Tip 4: Different motifs should be commensurately scaled

As is the case with orienting motifs consistently, scaling motifs commensurately makes the scene created by the tessellation more natural. For example, if giraffes and gorillas were used in a tessellation, it would look odd if they were the same height. In Figure 11.4, three different sea creatures, squids, rays, and sea turtles, are commensurately scaled.

Intentionally using a different scale for different motifs can create a whimsical effect, as shown in Figure 11.5. In this case, the design is more effective if the scales are very different. Note that monkeys with bananas and cats with mice are motifs that are naturally associated with each other.

Figure 11.4. The three motifs in this tessellation are scaled similarly to the size of actual squids, rays, and sea turtles.

Figure 11.5. One of the two motifs in each of these tessellations is much bigger relative to the other than it would be in real life, creating a whimsical effect.

Tip 5: Use source material to get the details right

A motif used for a tile can be something imaginary, like a dragon, or it can be something found in the real world. If a real-world motif like a lizard, fish, or flower is used, it will look more natural if it is modeled after a specific type of lizard, fish, or flower. Trying to draw motifs from memory is challenging and will probably not yield the best results. Identifying a real-world motif that looks similar to your tile can be very helpful in improving the appearance of the tile. You can find

source material like photographs and realistic illustrations in books or on the Internet, and these can help guide your design. You may want to place some photos or illustrations right next to your tile as you refine its shape and interior details.

For example, you might decide that your tile will be a fish with its tail folded over, or a flower or a ray, as shown in Figure 11.6. Initial attempts at adding details may not yield satisfactory results. You might realize it doesn't look right, but not know why. Looking at pictures of different types of fish, you might decide that your shape looks a little like a trout or a bass. You can then search images of these types of fish

Figure 11.6. Three different Escheresque tiles shown with no interior details, the first guess at interior details, and interior details informed by source material.

till you find one or two that you can use as models. You might find that you drew the eye too small or too high, or a fin is shaped wrong, or including the gills helps define the head. For the flower example, you might decide the petals look similar to those of a daffodil. Even though real daffodils have six petals, using a daffodil-like central structure makes for a reasonably realistic-looking flower. The manta ray provides another example, where the outline was modified slightly to accommodate more realistic anatomical details.

Tip 6: Stylize the design

While it's important to get the details right when designing tiles with real-world motifs, trying to include too much detail or trying to make it look too real are generally pitfalls to be avoided. The reason is that the constraints of making tiles fit together without gaps necessitate some distortion of motifs compared to their natural shapes. Trying to make a motif look too realistic tends to make these distortions more obvious. The optimum balance between stylizing a design and making it realistic will depend on how distorted the tile

is compared to the real object, as well as your rendering style.

The fact that every line segment is shared by two tiles means that the amount of convex (bulging outward from the center of the tile) and concave (curved inward toward the center of the tile) edges for the tiles in a tessellation must be equal. This means that it's impossible to create a tessellation consisting entirely of motifs that are completely made up of convex line segments. Most animals, such as frogs and people, consist mainly of convex surfaces, so a realistic drawing would need mostly convex lines. The drawings of Figure 11.7 contrast a realistic drawing of frogs with a tessellation of frogs. The feet are particularly challenging, and the sort of wide duck-like foot used here was commonly used by Escher in his frogs and lizards. The more realistic frogs consist almost entirely of convex lines, and it isn't possible to fit them together without gaps.

The cats of Figure 11.8 provide another example. In this case, the outline of the cat is rather abstract, consisting of straight-line segments, most of which would be convex curves in a more realistic drawing. Note that there are minimal interior details in the bodies, as more detail would just call attention to the distortion of the outlines.

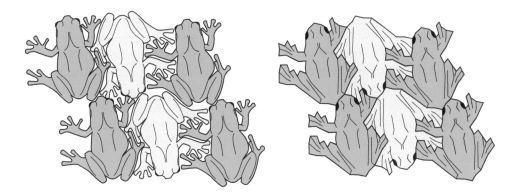

Figure 11.7. Natural-looking frogs are mostly made up of convex line segments, inevitably leading to gaps between them. Tessellating frogs are necessarily distorted to prevent gaps.

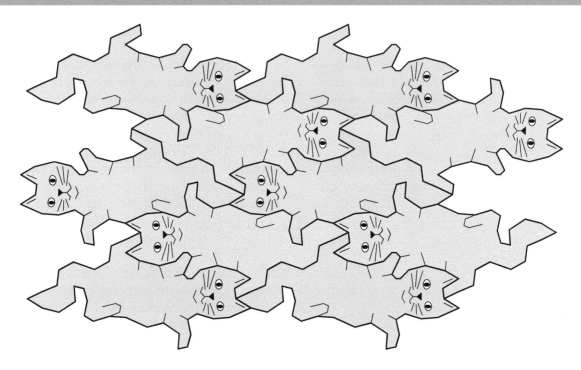

Figure 11.8. A tessellation of cats with short straight-line segments used both in the outline and interior details.

Tip 7: Choose a style that fits your taste and abilities

There are many options in the drawing style of a tessellation, just like any artwork, but some of these are unique to tessellations. A basic option discussed earlier in this chapter is whether to work by hand or by computer. Another option is the degree of realism vs. stylization. For a stylized drawing, there are options such as serious, cartoon-like, humorous, etc. Within these options, there are suboptions; e.g., if you choose a cartoon-like style, there are many options within it, from early Disney to contemporary Japanese manga. M.C. Escher's style in some cases tended toward the grotesque. This style includes fanciful or imagined creatures, so in that sense, it lent itself naturally to tessellations.

An aspect of style that is more or less unique to tessellations is the use or omission of border lines around tiles, since the borders define or delineate the tile boundaries. If border lines are used, they can be narrow or wide. M.C. Escher's borders varied from none to very wide. Coloring considerations, discussed in the next section, also uniquely impact tessellations, since there is no space between motifs.

The same tile is shown in Figure 11.9, with several different rendering styles. For all but the lower-right tile, the boundary and interior lines were created with a drawing program. In the upper left, a single solid color is used for the entire tile. In the upper right, the coloring was done by hand using colored pencils. In the middle left, two solid colors were used, while more details and colors were added at lower left. In the middle right, graduated coloring was done in Photoshop. The lower-right example uses photographic rendering,

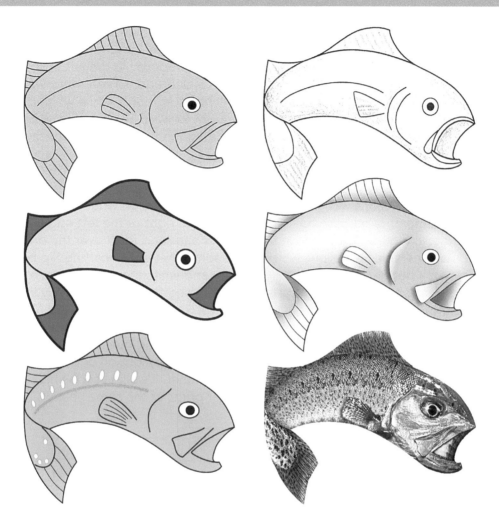

Figure 11.9. Six different ways of rendering the same fish tile.

where swatches of a photograph of an actual fish were distorted and blended in Photoshop to fill the tile boundary.

Tip 8: Choose colors that suit your taste and bring out the tiles

In a tessellation, it is usually desirable, but not essential, to use different colors for adjacent tiles to provide some contrast. Colors can be bold or subtle, similar or complementary, bright or muted, etc. You may want to draw from nature; e.g., the undersides of many rays are pale in color, while the tops are dark. Or you may wish to use less naturalistic colors. Each tile can be made essentially one color, or the color can be varied within a tile.

If it's compatible with your motifs, complementary colors are often a good choice. These are colors that are opposite to one another on the color wheel. In this case, I'm referring to an artist's color wheel, not a computer programmer's RGB color wheel. The complementary pairs of primary (red, blue, and yellow) and secondary (green, purple, and orange) colors are red-green, blue-orange, and purple-yellow. These colors naturally look good together and contrast nicely. If you want to use fairly realistic colors with a tessellation of real-world motifs, complementary colors may not be possible. For example, if you want to two-color a tessellation of lions, yellowish lions would look fine, but the complementary purplish lions would look strange. In Figure 11.10, I chose to use lighter and darker shades of yellow, which contrast less but look better overall.

Neighboring tiles can be greatly contrasting in value (lightness or darkness), or similar in value. One option for a two-colorable tessellation is to use black for one subset of tiles and white for the other tiles. This was used by Escher to achieve a sort of optical illusion in which figure (objects in the foreground) and ground (the background) reverse. Figure 11.11 illustrates figure/ground reversal in a tessellation of dragons. At the left end, the mind sees black dragons brought forward against a white background, while white dragons are brought forward against a black background at the right end. In the center region, where there is a tessellation of white and black dragons, whether the white dragons or black dragons are seen as being in the foreground depends largely on whether your eye was last on the right or the left edge. As the eye moves back and forth over the drawing, there is a somewhat disconcerting flip between black and white serving as a background. Escher used this trick effectively in one of his best-known prints, *Day and Night* (1938), which features a tessellation of black and white birds flying over a landscape in which a village is in daylight on one side and darkness on the other. It's also used in his two *Sky and Water* prints of the same year.

Figure 11.10. A lion tessellation, two-colored with complementary and more realistic colors.

Figure 11.11. "Dragon Metamorphosis", a digital print from 2003 that makes use of ground/figure reversal.

Activity 11.1. Finding motifs for a tile shape

Materials: Copies of Worksheet 11.1.

Objective: Learn to see motifs in tile shapes and to use interior lines to make the motif more convincing.

Vocabulary: Motif, interior details, orientation, viewpoint.

Activity Sequence

1. Write the vocabulary terms on the board and discuss the meaning of each.
2. Pass out copies of the worksheet.
3. For each tile, have the students imagine possible motifs by rotating the paper (looking at the tile in different orientations) and considering different viewpoints (e.g., side view or top view). Once they have a motif in mind, they should draw interior details.

Discussion Questions

1. For the tile shape a/b/c/d/e/f/g/h, what motif did you choose?
2. Did anyone choose a different motif? If so, what?
3. For the motif you chose, describe the orientation and viewpoint you chose and how you used interior details to make it look more like your motif.
4. How well do you think it turned out?
5. What might you have done differently?

Worksheet 11.1. Finding motifs for a tile shape

For each tile shape below, think of a real-world motif that could be used for it, and draw in interior details. It may help to rotate the sheet to look at the shapes in different orientations and to consider different viewpoints (e.g., top view vs. side view).

Activity 11.2.
Refining a tile shape using translation

Materials: Copies of Worksheet 11.2.

Objective: Learn to improve a tessellation that possesses translational symmetry by modifying the edges of the tile.

Vocabulary: Translation.

Activity Sequence

1. Write the vocabulary term on the board and discuss its meaning.
2. Pass out copies of the worksheet.
3. Have the students reshape the tile as described on the worksheet.

Discussion Questions

1. How did you reshape line a?
2. Do you think you succeeded in improving your motif? How?
3. How did you reshape line b?
4. Do you think you succeeded in improving your motif? How?

Worksheet 11.2. Refining a tile shape using translation

For the tile shape below, reshape the tile to improve your design from Worksheet 11.1. This tile has two distinct edges and fits together by translation. First reshape line a, then refine it, and then reshape line b and refine it. At each stage, the two instances of line a must be identical, and similarly for line b. Each time you reshape a line, redraw the interior details, modifying your earlier details to fit the new shape.

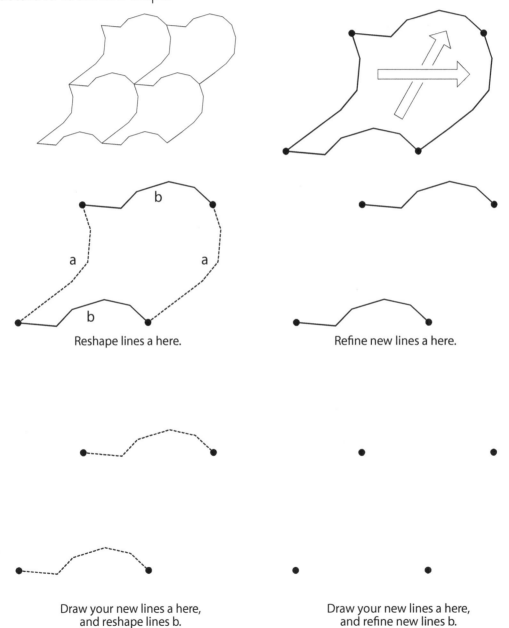

Reshape lines a here.

Refine new lines a here.

Draw your new lines a here,
and reshape lines b.

Draw your new lines a here,
and refine new lines b.

Activity 11.3. Refining a tile shape using glide reflection

Materials: Copies of Worksheet 11.3.

Objective: Learn to improve a tessellation that possesses glide reflection symmetry by modifying the edges of the tile.

Vocabulary: Glide reflection.

Activity Sequence

1. Write the vocabulary term on the board and discuss its meaning.
2. Pass out copies of the worksheet.
3. Have the students reshape the tile as described on the worksheet.

Discussion Questions

1. How did you reshape line a?
2. Do you think you succeeded in improving your motif? How?
3. How did you reshape line b? Did you find it difficult to draw two instances of the line, with the second instance mirrored compared to the first?
4. Do you think your modification of line b succeeded in improving your motif? How?

Worksheet 11.3. Refining a tile shape using glide reflection

For the tile shape below, reshape the tile to improve your design from Worksheet 11.1. This tile has two distinct edges and fits together laterally by translation and vertically by glide reflection. First reshape line a, then refine it, and then reshape line b and refine it. At each stage, the two instances of line b must be related by a reflection about a vertical line. This makes drawing the two instances of line b much trickier. Each time you reshape a line, redraw the interior details, modifying your earlier details to fit the new shape.

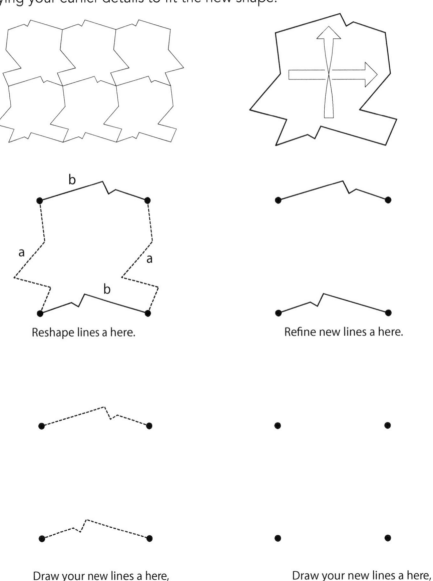

Reshape lines a here.

Refine new lines a here.

Draw your new lines a here, and reshape lines b.

Draw your new lines a here, and refine new lines b.

Activity 11.4.
Locating and using source material for real-life motifs

Materials: Access to a library or to a computer with Internet access. Copies of Worksheet 11.4.

Objective: Learn to locate source material using books or the Internet, and learn to use it to improve the appearance of the tiles for a tessellation.

Vocabulary: Source material, stylized drawing, realistic drawing.

Activity Sequence

1. Write the vocabulary terms on the board and discuss their meanings.
2. Pass out copies of the worksheet.
3. Have the students draw interior details in the leftmost outlines provided, working from their mental conception of what each motif should look like.
4. Have the students locate source material for each of the three motifs.
5. Have the students use their source material to draw interior details in the second and third copies of the outlines provided.

Discussion Questions

1. When you looked at the source material, were there things about the motifs that you hadn't realized? What were they, and how did they help your drawings?
2. Do you think your second or third drawings are better for each motif? Why?
3. Do you think your final drawing would look better if it were more stylized, or if it were more realistic? Why?

Worksheet 11.4. Locating and using source material for real-life motifs

For each tile shape below, first draw interior details in the leftmost outlines without using the source material. Then search in books or on the Internet for images of the motif given. Try to find at least three images for each motif that show it from a similar viewpoint to that shown below. Using those images to guide you, draw the interior details in the middle outline. Compare your drawing to the images you found and decide how your drawing could be improved. Consider the degree to which the details are stylized vs. being realistic. Then draw interior details again in the rightmost outline.

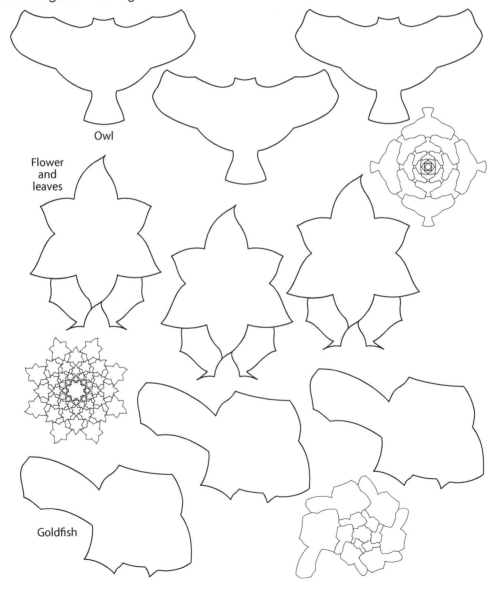

Owl

Flower and leaves

Goldfish

Chapter 12

Special Techniques to Solve Design Problems

Producing a really good tessellation generally takes hours of refinement. If you get stuck, do something else for a while and then come back to it. Coming back to a design after taking a break for hours or even days allows you a fresh and more objective look at it, and you will usually see a way to improve it. In writing this book, I looked back at the tessellations I had designed years earlier and saw ways to improve many of them. Figure 12.1 shows some steps in a seahorse tessellation that I first designed in 1992, wasn't satisfied with, and came back to it a few times over the years till I made the final version in 2008.

Sometimes you just can't get a tile to look right with the motif and the starting geometric tiling you've chosen. There are some special tricks that can help you out in such cases. Four of these techniques are described in this chapter, along with examples.

Figure 12.1. A first draft of a seahorse tessellation (left), with two later refinements of the motif (center), and a final version (right).

Technique 1: Distorting the entire tessellation

An entire tessellation can be stretched or skewed to change the shape of the tiles. This can be used to improve the appearance of Escheresque tiles. For example, if you create a tessellation of rabbits, and they look too short and squat, you can stretch them in the vertical direction (Figure 12.2). If the tessellation

Figure 12.2. Stretching an entire tessellation to change the shape of a motif.

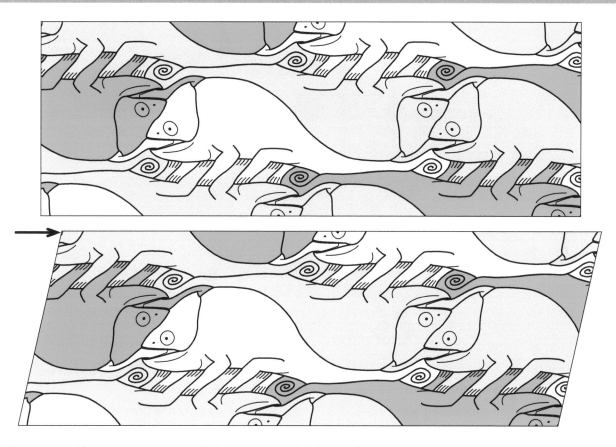

Figure 12.3. Skewing an entire tessellation to change the shape of the motif.

motif appears in multiple orientations, stretching might not work as it affects the different orientations differently.

Similarly, skewing a tessellation (pushing one side, while leaving the opposing side fixed) can be used to modify the appearance of the tiles, as shown with the chameleon tessellation of Figure 12.3. The chameleons appear in two different orientations, but skewing the tessellation affects each orientation the same way. This wouldn't be true for many tessellations. One of the advantages of working on a computer is the ease with which distortions of this sort can be carried out.

Technique 2: Breaking symmetries

For some motifs, a highly symmetric drawing may not allow tiles to fit together without gaps. Making a basically symmetric object not quite symmetric can solve this problem and make the tiles look more interesting as well. An example is the pteranodon tile and tessellation of Figure 12.4. A pteranodon, like many living things, possesses bilateral symmetry. Drawn symmetrically, the beak and tail get in the way of one another. Tilting each to the side a bit allows them to fit together without leaving gaps

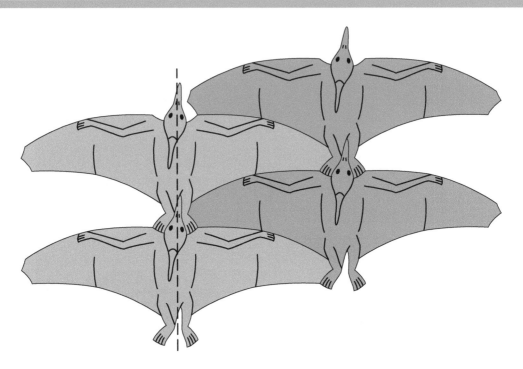

Figure 12.4. A tessellation of pteranodons in which the mirror symmetry was broken by shifting the head and tail, allowing the tiles to fit together. The dashed line is the center line for the wings and legs.

or having to use an excessively short beak. Arriving at these pteranodon tiles from a simple starting tessellation of rhombi involves moving and splitting vertices, described in Technique 4.

Another example is provided by the flower tessellation of Figure 12.5. The flowers on the left have six-fold rotational symmetry, and each petal has bilateral symmetry. They

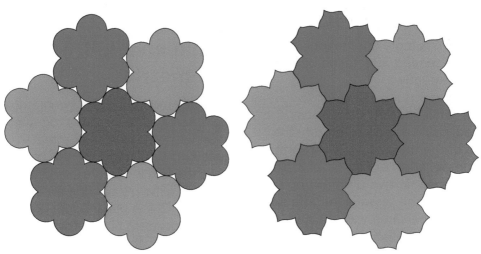

Figure 12.5. Flowers that don't quite fit together can be modified to fit by breaking the mirror symmetry of the motif.

nearly fit together, but small gaps remain between them. Modifying the petals in a way that removes their bilateral symmetry allows the gaps to be filled so the flowers tessellate. Note that the six-fold rotational symmetry is maintained.

Technique 3: Splitting a tile into smaller tiles

Sometimes you want to use a particular tile but can't figure out a good motif for it. The tiling of Figure 12.6 has a unit cell of eight tiles, outlined with a heavy line, in which each of the four different tiles appears two times. On the left, three of the four have had interior details added to form frogs, toads, and snakes. The last tile could be made into another snake, but the shape doesn't lend itself very well to a snake. A better solution is to divide it in two, after which interior details

can be added to form two tadpoles, as shown on the right. Escher used this technique in several of his tessellations. In the example here, the tile is divided into two smaller tiles of the same type. When Escher used this technique, he generally divided a tile into two different tiles.

Technique 4: Splitting and moving vertices

A more sophisticated and very effective way of reshaping tiles in order to improve their shapes is to split and move vertices. This way, you are not constrained to rigidly conform to your starting geometric template.

As an example, consider the pteranodon tessellation at the top in Figure 12.7. Let's say you aren't satisfied with the motif and wish you had more freedom in shaping the beak and back feet. You can achieve this by

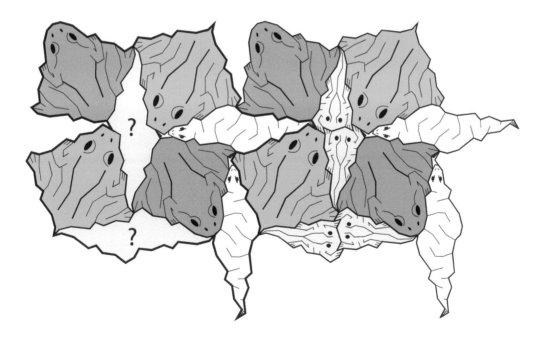

Figure 12.6. A tessellation in which a tile that proved difficult to be made into a reptile motif was split in two, allowing the formation of a pair of tadpoles.

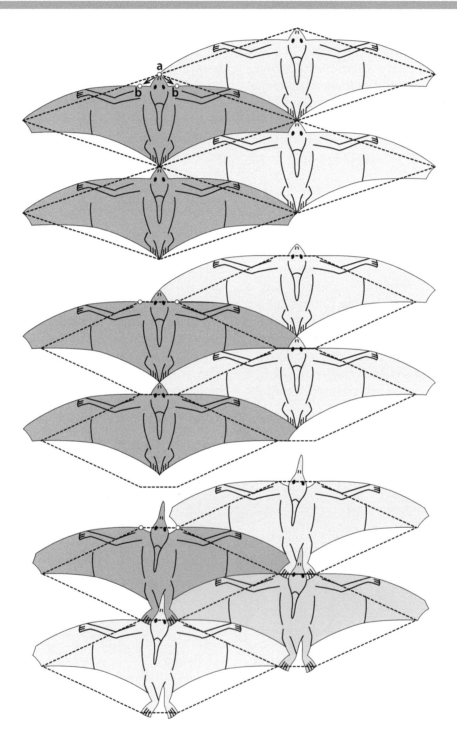

Figure 12.7. Splitting and moving the vertices in the top tessellation from point (a) to point (b) allowed the motif to be reshaped to allow a more realistic beak.

splitting and moving the vertex marked "a" so that it is replaced with the two vertices marked "b". Note that each tile meets four vertices of the tessellation, all of the same type. This means that every vertex will split and move the same way.

The middle drawing shows how this affects the geometric tessellation underlying the design. Each rhombus is replaced by a hexagon, and each tile now meets six vertices instead of four. You are now free to reshape the horizontal lines between the two new vertices in any way you wish. In the bottom drawing, these have been reshaped to allow a longer beak and feet that extend out. The other edges were modified as well, creating a more realistic pteranodon. Note that there are now two types of vertices, one meeting a left foot and one meeting a right foot.

Splitting vertices can change the coloring of a tessellation. In the example of Figure 12.7, the starting tessellation was two-colorable, while the new tessellation requires three colors if no adjacent tiles are of the same color.

Activity 12.1. Reshaping a tile by splitting and moving vertices

Materials: Copies of Worksheet 12.1.
Objective: Learn to improve the shape of an Escheresque tile by splitting and moving vertices of the tessellation.
Vocabulary: Vertex (plural form vertices).
Activity Sequence
1. Write the vocabulary term on the board and review its meaning.
2. Pass out copies of the worksheet.
3. Go over the top portion of the worksheet, making sure the students understand step 1.
4. Have the students redraw the tile as directed in step 2, with the side vertices split and moved. Be sure that all of the students modify the tile correctly so that it tessellates.
5. Have the students add interior details.

Discussion Questions
1. Did you find it difficult to complete the reshaping of the tile in step 2? Why or why not?
2. How exactly did you reshape the edges? Do you think your design is effective? Why or why not? Did anyone modify the edges differently? Compare and contrast the different approaches.
3. How did you change the interior lines to improve the motif? Which of the various changes made by different students worked the best?

Worksheet 12.1. Reshaping a tile by splitting and moving vertices

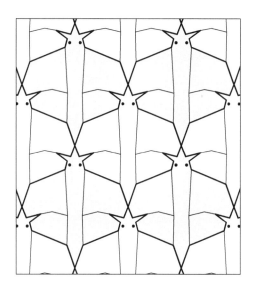

Suppose you made the butterfly tessellation shown in Figure 12.8 but weren't satisfied with it. Perhaps you think it could be improved if there were a way to shorten the body and make the wings bigger. The steps below will walk you through using the technique of splitting and moving vertices to achieve this goal.

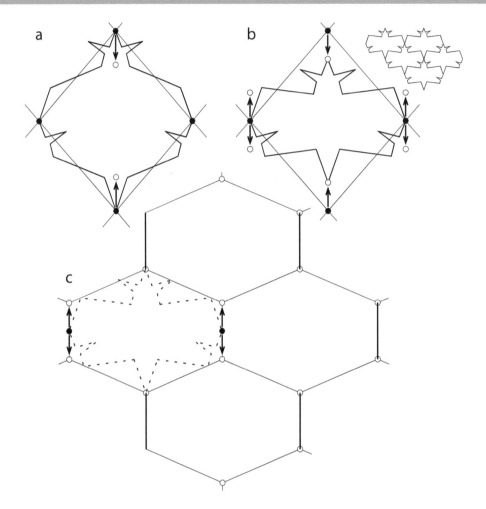

Step 1. The butterfly tile is based on a tessellation of rhombi. Each tile meets four vertices of the tessellation, all of the same type. Suppose you decide to try moving the top vertex down and the bottom vertex up, up by the same amount, as shown in (a). This reshapes the tile, as shown in (b), making the body shorter. The wings are too small now, though.

Step 2. Note that the underlying geometric template is now a tessellation of hexagons (c). Redraw the angled lines that connect the vertices at the top and the bottom of the vertical lines. Note that each of these lines will have the same shape, but half will be reflected.

Step 3. Now draw new interior lines to improve the appearance of the motif; e.g., you might want to make the body narrower.

Chapter 13

Escheresque Tessellations Based on Squares

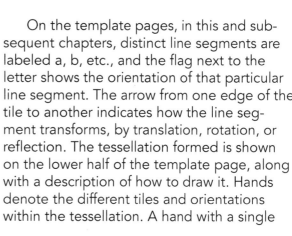

A grid of squares is the simplest geometric basis for a tessellation and therefore a good starting point in learning to design tessellations. At the same time, a square grid offers a wide range of possibilities, including tessellations with translational, rotational, and glide reflection symmetries.

In this chapter, templates are presented that provide examples of each type of symmetry. Square grids are provided that can be used to guide the design of tessellations using hand drawing and cutting techniques. Every template and example in this chapter uses a square grid of this scale. A detailed step-by-step example shows how to use the templates and grids to create your own tessellation designs. The same general procedure can be used with the templates in later chapters.

On the template pages, in this and subsequent chapters, distinct line segments are labeled a, b, etc., and the flag next to the letter shows the orientation of that particular line segment. The arrow from one edge of the tile to another indicates how the line segment transforms, by translation, rotation, or reflection. The tessellation formed is shown on the lower half of the template page, along with a description of how to draw it. Hands denote the different tiles and orientations within the tessellation. A hand with a single

Template

upheld finger indicates the first prototile of the tessellation, while two fingers indicate the second prototile. The orientation of the hands indicates the orientation of the tiles, while Roman numerals next to the hands enumerate distinct orientations of the tile.

The templates can be used to create tessellations either by hand or using a drawing program on a computer. The detailed example in this chapter shows how to create a tessellation by hand, but it is straightforward to mimic these steps using a computer drawing program.

M.C. Escher utilized square grids in many of his tessellations. In his famous 1939–1940 woodcut *Metamorphose II*, lizards emerge from a checkerboard pattern of black and white squares. Other motifs he formed using square grids include ants, angels and devils (two of each per square), and fish.

While a square grid is the easiest geometric basis to work with, from a graphic design standpoint, square grids can lead to designs that are blocky or static, lacking motion and life. This is particularly true in a tessellation that only possesses translational symmetry. This doesn't have to be the case, but it's something to be aware of and to consciously avoid.

A closely related geometric basis is a grid of rectangles. These are not treated as a separate topic in this book, but many of the square templates presented in this chapter and the next can be made into rectangular templates simply by stretching the design in one direction, as described previously.

Creating a tessellation by hand

This detailed example uses the square grids and Template 13.3 found later in this chapter. It illustrates a general method for designing tessellations by hand, using the various templates in this book.

Step 1. Copy the Square Tile Grids (Figure 13.4) onto a heavy paper, such as card stock. Copy the Square Tessellation Grid (Figure 13.5) onto a lighter paper that works well for drawing and coloring.

Step 2. For the template chosen, mark the appropriate letters and flags showing the relationship between the various sides of the shape (a square in this case). Do this for each of the tile grids. Note that for some templates, you will have two different tiles. In this case, you may want to draw each one on a separate tile grid.

Step 3. Using these markings as a guide, draw curves (these can consist of straight-line segments) that follow the rules indicated by the letters and flags. The large arrows on the template sheets show how a line segment moves from one edge to another. It is probably easiest to start by connecting crossing points on the grid with straight-line segments, as shown in Figure 13.1a.

Step 4. If your shape suggests a motif that you wish to develop further, refine the shape of your curves on the same or a fresh tile grid. You may wish to rough in the interior details at this point, as shown in Figure 13.1b. If you don't like the shape (your tile), you may wish to try all new starting curves.

Step 5. Once you are satisfied with the tentative outline, refine it further and draw in interior details, as shown in Figure 13.1c. This is what each tile in your tessellation will look like if you only have one type of tile. Keep the orientation of the tiles in the final tessellation in mind. For example, for the template chosen here, half the fish would be upside down. Having upside-down fish can be avoided by drawing different interior details for those tiles, as shown in Figure 13.1d, where the upside-down tile has been split into two smaller tiles

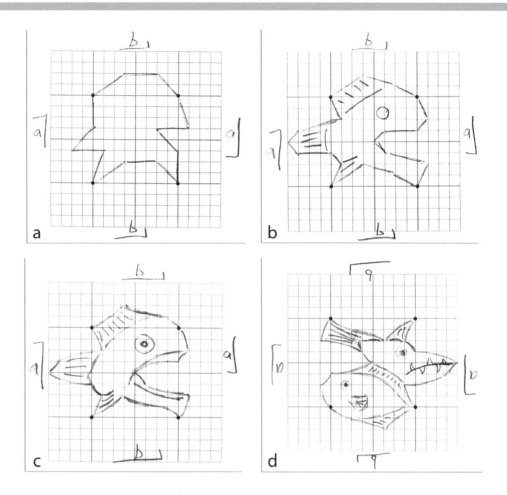

Figure 13.1. First steps in designing an Escheresque tile by hand.

Step 6. Once you're satisfied with your tile, carefully cut curves as instructed for your particular template. In the example here, cut one of the two "a" curves and one of the two "b" curves. Cut off the bulk of the sheet, so you have a piece similar to that shown in Figure 13.2a, that can be used as a tracing template.

Step 7. Use this template to draw a tile on the appropriate tessellation grid sheet. Carefully align the shape to the grid using the dots. Hold it firmly in place and tightly trace the cut curves, as shown in Figure 13.2b. It's important to use a sharp pencil and trace right at the edge of the template. Then reposition the template to trace the other curves. The template page shows how the tiles fit together and how each tile is oriented. Use it as a guide to ensure the curves are traced in the correct orientations. Template 13.3 shows that both the "a" and "b" curves are oriented the same within a column, but that they are mirrored from one column to the adjacent column. This means the template needs to be flipped over to draw these curves, as shown in Figure 13.2c.

Step 8. Repeat this process of aligning the template with dots and tracing the template until the entire tessellation grid sheet (or as much as you want) is covered with your tessellated shapes.

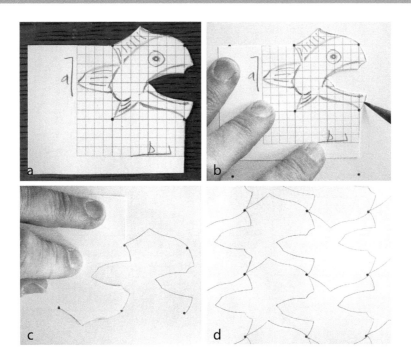

Figure 13.2. Intermediate steps in designing an Escheresque tile by hand.

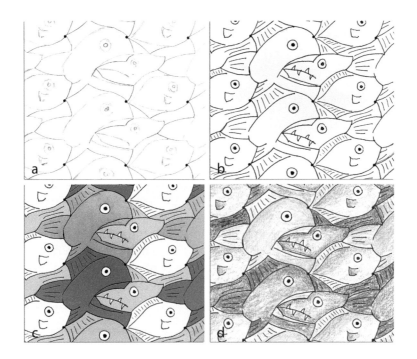

Figure 13.3. Final steps in designing an Escheresque tile by hand.

Square Tile Grids

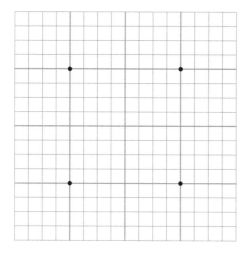 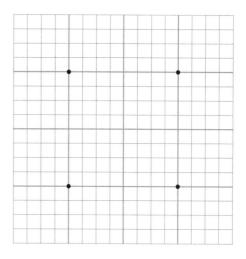

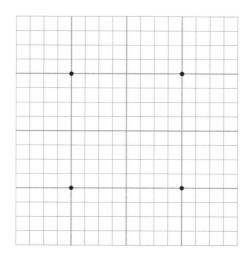 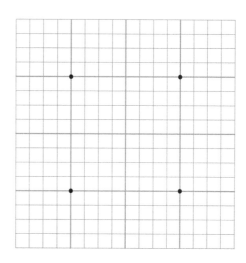

Figure 13.4. Grids to use for designing a tile for a square-based tessellation.

Step 9. Lightly draw in the key interior lines using a pencil, as shown in Figure 13.3a.

Step 10. Draw the final lines. You may wish to use an ink pen for this step. This will allow the pencil marks to be erased once the ink is fully dry, as shown in Figure 13.3b. Otherwise, smearing of pencil marks could occur during coloring.

Step 11. Color your tessellation. This can be done using crayons, colored pencils, felt-tip markers, etc. The tiles can be colored uniformly (Figure 13.3c, where the filling was done in Photoshop after scanning the hand-drawn tiles), or shaded to give them more of a three-dimensional appearance (Figure 13.3d). It's usually

Square Tessellation Grid

Figure 13.5. Grid to use for drawing a square-based tessellation.

a good idea to use different colors for adjacent tiles, so they stand out better. Choose whatever colors you like for your particular tiles. You may wish to make multiple copies of your tessellation after Step 10, so you can experiment with different colors without having to redraw the tiles.

Use the grids in Figures 13.4 and 13.5 to design tiles using the templates found later in the chapter. First choose a template, and then copy the letters and flags for each line segment on a Square Tile Grid (Figure 13.4). Then draw curves (these can consist of straight-line segments) connecting the dots in a manner that follows the rules described by the letters and flags. Make sure the curves won't cross anywhere in the full tessellation. If you find a tile you like, refine the shape of the curves and add interior details to create a motif for your tessellation. Then cut out your tile as directed for the template you're using, and follow the directions for that template to

form a tessellation on a copy of the Square Tessellation Grid (Figure 13.5).

Template 13.1. Tessellation with translational symmetry only

This template, shown in Figure 13.6, has the symmetry group *p*1 and the Heesch type TTTT. Escher's tessellations nos. 105, 106, 127, and 128 [Schattscheider 2004] are based on this template. This is the simplest way to form a tessellation using a square grid. There are two independent line segments, each of which simply translates from one side of the square to the other.

To draw this tessellation using the grids, cut along the top and right edges of your tile. Form Line 1, as indicated in Figure 13.7, by tracing the cut edges. Repeatedly draw this same line in the same orientation to form a larger tessellation.

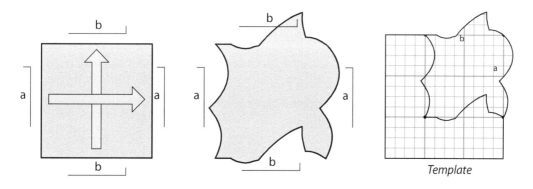

Figure 13.6. Template for a square-based Escheresque tile that admits a tessellation by simple translation.

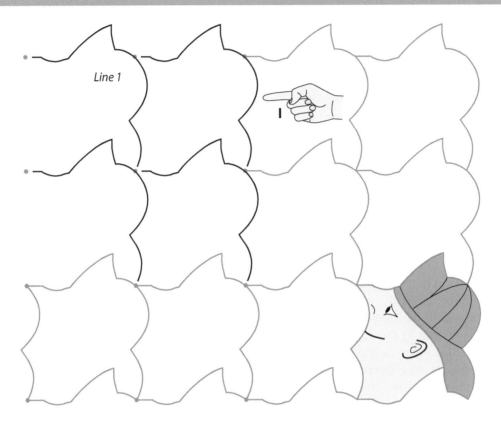

Figure 13.7. Template for drawing a tessellation using a tile designed according to the template of Figure 13.6.

Template 13.2. Tessellation with two- and four-fold rotational symmetry

This is another simple template, but one that forms a more interesting tessellation, with both two-fold and four-fold rotational symmetry (Figure 13.8). It has the symmetry group *p4* and the Heesch type $C_4C_4C_4C_4$. Escher's tessellations nos. 14, 15, 20, 23, and 104 are based on this template. The tile appears in four different orientations in the tessellation.

To draw this tessellation using the grids, cut along the top and left edges of your tile.

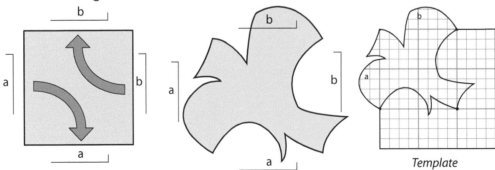

Figure 13.8. Template for a square-based Escheresque tile that admits a tessellation with rotational symmetry.

Figure 13.9. Template for drawing a tessellation using a tile designed according to the template of Figure 13.8.

Form Line 1, as indicated in Figure 13.9, by tracing the cut edges. Then rotate your template to 90° clockwise and form Line 2. Rotate clockwise to 90° again to form Line 3 and one last time to form Line 4. Repeat this group of four lines to form a larger tessellation.

The tessellation of Figure 13.10 is based on Template 13.2. In this design, each tile has been divided into two smaller tiles. Using the squares to guide you, try drawing additional tiles on the top row. Then add details to these and the blank tiles in the figure. You can also color the tessellation or mark the points of rotational symmetry.

Figure 13.10. A tessellation of seahorses and eels based on Template 13.2.

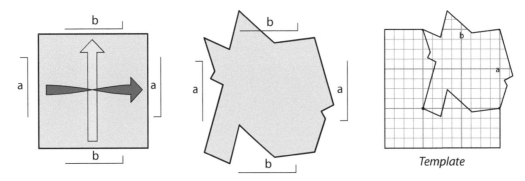

Template

Figure 13.11. Template for a square-based Escheresque tile that admits a tessellation with glide reflection symmetry.

Template 13.3. Tessellation with glide reflection symmetry

This template, shown in Figure 13.11, has the symmetry group *pg* and the Heesch type TGTG. One of the two independent line segments simply translates from one side of the square to the opposite edge, while the other line segment is reflected from one side of the square to the other. This results in a tessellation possessing glide reflection symmetry, in which the tile appears in two distinct orientations. Escher's tessellations nos. 108 and 109 are based on this template.

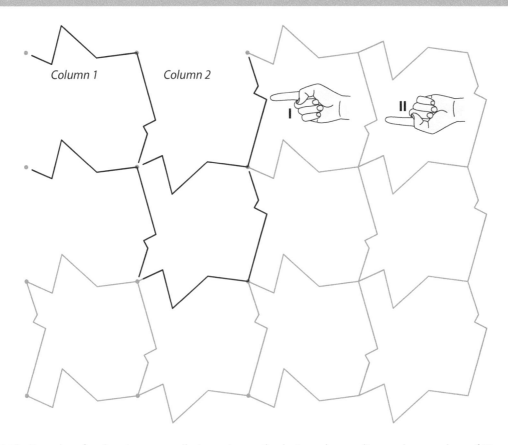

Column 1   Column 2

I   II

Figure 13.12. Template for drawing a tessellation using a tile designed according to the template of Figure 13.11.

To draw this tessellation using the grids, cut along the top and right edges of your tile. Form Column 1 of lines, as indicated in Figure 13.12, by tracing the cut edges. Then flip your template over from top to bottom and form an adjacent Column 2 of lines. Repeat this pattern to make a wider tessellation.

Template 13.4. Tessellation with a simple reflection and glide reflection symmetry

This template, shown in Figure 13.13, only has one independent line segment, but multiple lines of glide reflection symmetry, as well as simple reflection lines and rotational symmetry. The motif, which possesses bilateral symmetry, appears in four different orientations. The symmetry group is *p4g*. Recall that tiles with bilateral symmetry are excluded from Heesch's classification system. Escher's tessellations nos. 13, 86, 122, and 125 are based on this template.

To draw this tessellation using the grids, cut along the top edge of your tile. Form Line 1, as indicated in Figure 13.14, by tracing the cut edge. Then rotate your template to 90° counter-clockwise to form Line 2. Next flip it over from top to bottom to trace Line 3. Finally rotate it to 90° clockwise to form Line 4. Form Lines 5–8 as shown, and then repeat this group of lines to form a larger tessellation.

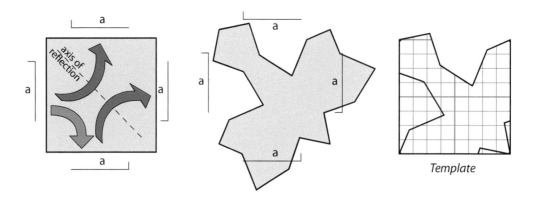

Figure 13.13. Template for a square-based Escheresque tile that admits a tessellation with glide reflection and simple reflection symmetry.

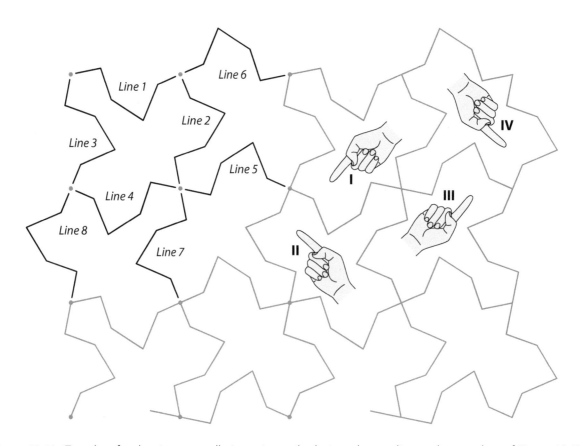

Figure 13.14. Template for drawing a tessellation using a tile designed according to the template of Figure 13.13.

Template 13.5. Tessellation with glide reflection symmetry in two orthogonal directions

This template, shown in Figure 13.15, has the symmetry group *pgg* and the Heesch type $G_1G_2G_1G_2$. Each of two distinct line segments is reflected from one side of the square to the other, resulting in a tessellation possessing both two-fold rotational symmetry and two distinct orthogonal lines of glide reflection symmetry. The tile appears in four orientations.

To draw this tessellation using the grids, cut along the top and right edges of your tile. Form Line 1, as indicated in Figure 13.16, by tracing the cut edges. Then flip your template over from right to left and form Line 2. Next flip it over from top to bottom and form Line 3. Then flip it over from right to left and form Line 4. Repeat this group of four lines to form a larger tessellation.

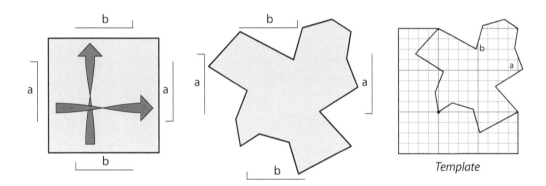

Figure 13.15. Template for a square-based Escheresque tile that admits a tessellation with glide reflection symmetry in two orthogonal directions.

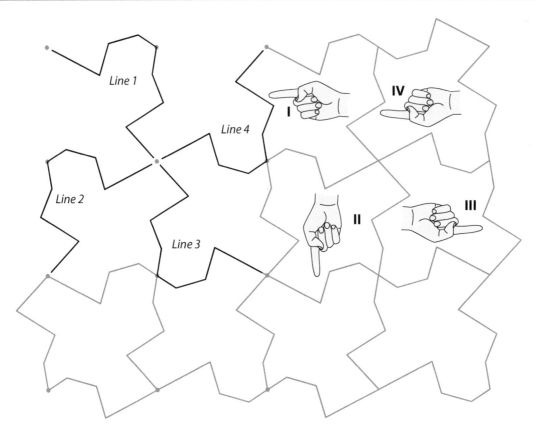

Figure 13.16. Template for drawing a tessellation using a tile designed according to the template of Figure 13.15.

Template 13.6. Tessellation with two-fold rotational and glide reflection symmetry

This template, shown in Figure 13.17, has the symmetry group *pmg*. This template contains a bilaterally symmetric tile that occurs in two different orientations. There are two independent line segments, each of which occurs four times in a tile. Escher's tessellations nos. 37 and 89 are based on this template.

To draw this tessellation using the grids, cut along the right half of the top edge and the top half of the right edge of your tile. Form Line 1, as indicated in Figure 13.18, by tracing along the cut edges. Then rotate your template to 180° and form Line 2. Next, flip it over about a diagonal line running from the top left to bottom right corners of the tile and form Line 3. Finally, rotate it to 180° and form Line 4. Repeat this group of four lines to form the larger tessellation.

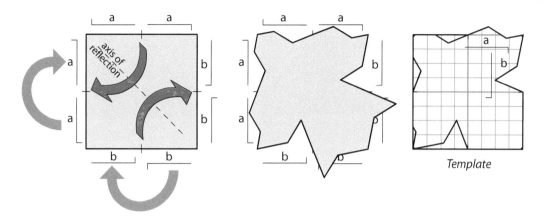

Figure 13.17. Template for a square-based Escheresque tile that admits a tessellation with two-fold rotational and glide reflection symmetry.

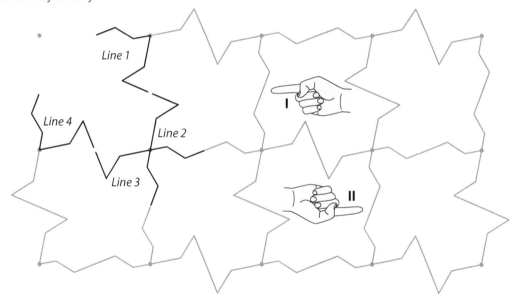

Figure 13.18. Template for drawing a tessellation using a tile designed according to the template of Figure 13.17.

Template 13.7. Tessellation with two motifs and glide reflection symmetry

The template, shown in Figure 13.19, has the symmetry group *pg*. This template has two motifs, each of which occurs in two orientations and has glide reflection symmetry. This is the same template used for the cats and mice design earlier in the book, except it is stretched vertically for that design.

To draw this tessellation using the grids, cut along the left, top, and right edges of your first tile. Form Line 1, as indicated in Figure 13.20, by tracing along the cut edges. Then slide it over and trace the top line only to form Line 2. Next, flip it over from left to right and form Line 3. Finally, slide it over and trace the top edge only to form Line 4. Repeat this group of four lines to form a larger tessellation.

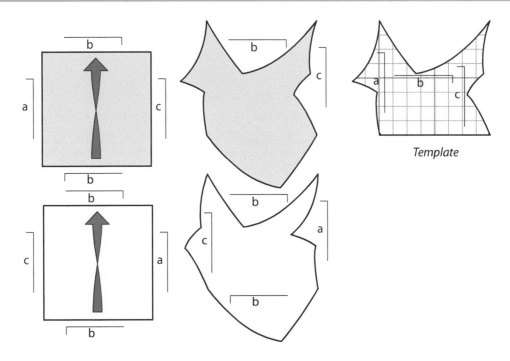

Figure 13.19. Template for a pair of square-based Escheresque tiles that together admits a tessellation with glide reflection symmetry.

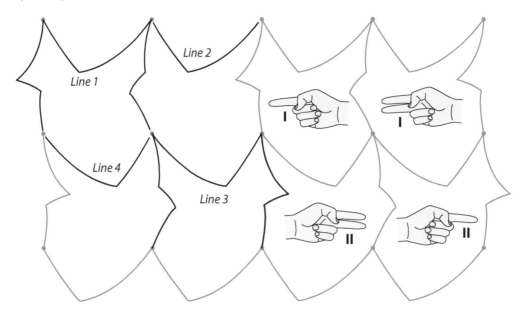

Figure 13.20. Template for drawing a tessellation using a pair of tiles designed according to the template of Figure 13.19.

Template 13.8. Tessellation with two different tiles and reflection symmetry in one direction

The template, shown in Figure 13.21, has the symmetry group *pm*. This template contains two different tiles, each of which possesses reflection symmetry about a center line. This sort of symmetry lends itself well to living creatures seen from top or front views.

To draw this tessellation using the grids, completely cut out your top tile. Form Line 1 as indicated in Figure 13.22 by tracing all the way around your template. Translate your template and trace completely around it to form Line 2. Repeat to form a larger tessellation.

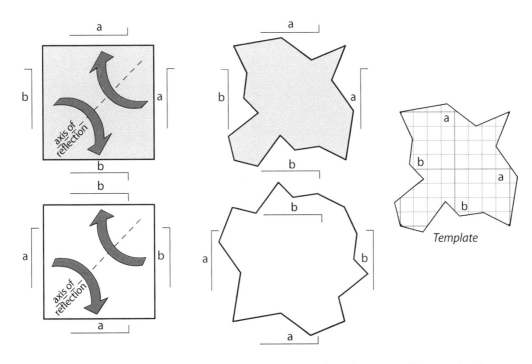

Figure 13.21. Template for a pair of square-based Escheresque tiles that admit a tessellation with reflection symmetry.

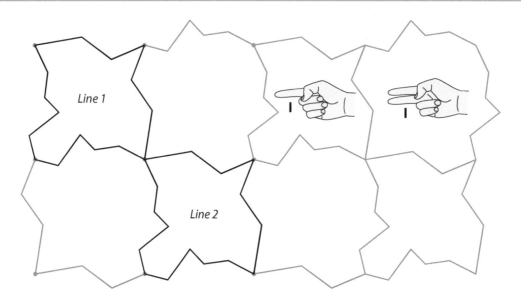

Figure 13.22. Template for drawing a tessellation using a pair of tiles designed according to the template of Figure 13.21.

Template 13.9. Tessellation with two different tiles and reflection symmetry in two orthogonal directions

This template, shown in Figure 13.23, has the symmetry group *p4m*. This template contains two different tiles. The tessellation contains multiple lines of reflection in orthogonal directions, as well as points of two- and four-fold rotational symmetry. A single line segment appears eight times in each tile.

To draw this tessellation using the grids, cut along the right half of the top edge and the top half of the right edge of your first tile. Form Line 1, as indicated in Figure 13.24, by tracing along the cut edges. Then rotate it to 90° clockwise to form Line 2 and again for Lines 3 and 4. Repeat this set of four lines to form a larger tessellation.

The flower design of Figure 13.25 is based on Template 13.9. Try finishing the top two rows of the tessellation. Where are the lines of glide reflection symmetry (ignoring small differences in interior details) and points of rotational symmetry?

The steps in the activities are described in more detail in the pages on creating a tessellation by hand. Unless otherwise noted, a pencil should be used. In a classroom setting, the teacher may wish to demonstrate the steps before having the students perform them.

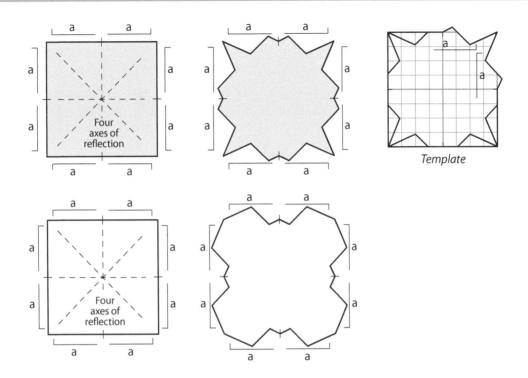

Figure 13.23. Template for a square-based pair of Escheresque tiles that admit a tessellation with reflection lines in orthogonal directions.

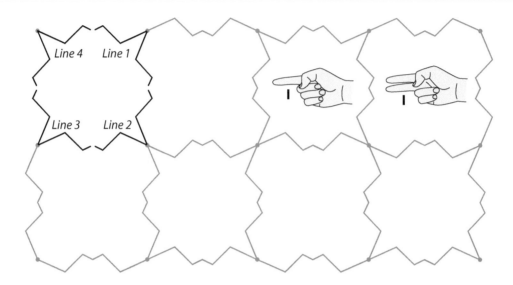

Figure 13.24. Template for drawing a tessellation using a pair of tiles designed according to the template of Figure 13.23.

Figure 13.25. A tessellation of flowers based on Template 13.9.

Activity 13.1. Creating an Escheresque tessellation with translational symmetry

Materials: Copies of template Figures 13.6 and 13.7 on paper, copies of the Square Tile Grids (Figure 13.4) sheet on card stock, copies of the Square Tessellation Grid (Figure 13.5) on a paper that can be drawn and colored on, scissors or craft knife, pencil, pen, eraser, and crayons or colored pencils (or other means of coloring a tessellation).

Objective: Learn to create, by hand, a tessellation that possesses translational symmetry.

Vocabulary: Translational symmetry, template, vector, tile, prototile, motif.

Activity Sequence

1. Write the vocabulary words on the board and discuss the meaning of each one.
2. Pass out the copies and other materials.
3. On Figure 13.7, have the students draw two different vectors indicating translation distances and directions that would cause the tessellation to perfectly overlap itself.
4. Have the students mark the letters and flags for each of the four Square Tile Grids in a similar fashion to how they're marked on the template sheet.
5. Using the letters and flags as a guide, have them draw curves connecting the grid dots on at least one of the grids. The two "a" lines should be identical, as should the two "b" lines.
6. Have the students identify a motif in (one of) their shape(s). Then have them refine the shape and rough in the interior details.
7. Have them further refine the shape and interior details to produce a prototile.
8. Have them cut out one curve of each type (one for each letter) and cut off the bulk of the sheet to create a template, as shown in Figure 13.6.
9. Have them use their templates to draw the outlines of the tiles on the Square Tessellation Grid sheet, as directed for Template 13.1, continuing until the sheet is filled up.
10. Have them lightly sketch in key interior details for each tile.
11. Have the students use an ink pen to go over the tile outlines and interior details for the entire tessellation. After the ink is fully dry, unwanted pencil marks can be erased.
12. Have them color their tessellations. Before starting, you may wish to discuss coloring options.

Discussion Questions

1. What motif did you use for your tile? Do you think it was effective? How would you change it if you could do it over again?
2. What step did you find the most challenging? Why?

Activity 13.2. Creating an Escheresque tessellation with rotational symmetry

Materials: Copies of template Figures 13.8 and 13.9 on paper, copies of the Square Tile Grids (Figure 13.4) sheet on card stock, copies of the Square Tessellation Grid (Figure 13.5) on a paper that can be drawn and colored on, scissors or craft knife, pencil, pen, eraser, and crayons or colored pencils (or other means of coloring a tessellation).

Objective: Learn to create, by hand, a tessellation that possesses rotational symmetry.

Vocabulary: Rotational symmetry.

Activity Sequence

1. Write the vocabulary term on the board and discuss its meaning.
2. Pass out the copies and other materials.
3. On Figure 13.9, have the students mark each distinct point of rotational symmetry, using a polygon to indicate the amount of rotation. For two-fold rotation, use a rectangle. For four-fold rotation, use a square.
4. Have the students mark the letters and flags for each of the four Square Tile Grids in a similar fashion to how they're marked on the template sheet.
5. Using the letters and flags as a guide, have them draw curves connecting the grid dots on at least one of the grids. The two "a" lines should be identical, as should the two "b" lines.
6. Have the students identify a motif in (one of) their shape(s). Then have them refine the shape and rough in the interior details.
7. Have them further refine the shape and interior details to produce a prototile.
8. Have them cut out one curve of each type (one for each letter) and cut off the bulk of the sheet to create a template, as shown in Figure 13.8.
9. Have them use their templates to draw the outlines of the tiles on the Square Tessellation Grid sheet, as directed for Template 13.2, continuing until the sheet is filled up.
10. Have them lightly sketch in key interior details for each tile.
11. Have the students use an ink pen to go over the tile outlines and interior details for the entire tessellation. After the ink is fully dry, unwanted pencil marks can be erased.
12. Have them color their tessellations. Before starting, you may wish to discuss coloring options.

Discussion Questions

1. What motif did you use for your tile? Do you think it was effective? How would you change it if you could do it over again?
2. What step did you find the most challenging? Why?
3. Did you find designing this tessellation to be more or less difficult than designing the tessellation in Activity 13.1? Why or why not?

Activity 13.3. Creating an Escheresque tessellation with glide reflection symmetry

Materials: Copies of template Figures 13.11 and 13.12 on paper, copies of the Square Tile Grids (Figure 13.4) sheet on card stock, copies of the Square Tessellation Grid (Figure 13.5) on a paper that can be drawn and colored on, scissors or craft knife, pencil, pen, eraser, and crayons or colored pencils (or other means of coloring a tessellation).

Objective: Learn to create, by hand, a tessellation that possesses glide reflection symmetry.

Vocabulary: Glide reflection symmetry.

Activity Sequence
1. Write the vocabulary term on the board and discuss its meaning.
2. Pass out the copies and other materials.
3. On Figure 13.12, have the students mark the lines of glide reflection symmetry, along with the glide vectors that cause the tessellation to perfectly overlie itself.
4. Have the students mark the letters and flags for each of the four Square Tile Grids in a similar fashion to how they're marked on the template sheet.
5. Using the letters and flags as a guide, have them draw curves connecting the grid dots on at least one of the grids. The two "b" lines should be identical, while the two "a" lines should be mirror images of each other.

6. Have the students identify a motif in (one of) their shape(s). Then have them refine the shape and rough in the interior details.
7. Have them further refine the shape and interior details to produce a prototile.
8. Have them cut out one curve of each type (one for each letter) and cut off the bulk of the sheet to create a template, as shown in Figure 13.11.
9. Have them use their templates to draw the outlines of the tiles on the Square Tessellation Grid sheet, as directed for Template 13.3, continuing until the sheet is filled up.
10. Have them lightly sketch in key interior details for each tile.
11. Have the students use an ink pen to go over the tile outlines and interior details for the entire tessellation. After the ink is fully dry, unwanted pencil marks can be erased.
12. Have them color their tessellations. Before starting, you may wish to discuss coloring options.

Discussion Questions
1. What motif did you use for your tile? Do you think it was effective? How would you change it if you could do it over again?
2. What step did you find the most challenging? Why?
3. Did you find designing this tessellation to be more or less difficult than designing the tessellations in Activities 13.1 and 13.2? Why or why not?
4. Which of the three templates, 13.1, 13.2, or 13.3, is your favorite? Why?

Chapter 14

Escheresque Tessellations Based on Isosceles Right Triangle and Kite-Shaped Tiles

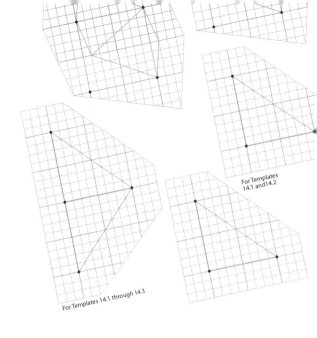

For Templates 14.1 and 14.2

For Templates 14.1 through 14.3

I n this chapter some tessellations based on isosceles right triangles and particular kite-shaped tiles are described. These can be drawn on the same Square Tessellation Grid in Figure 13.5. Additional tile grids tailored to these shapes are provided in Figure 14.1.

The right triangles that are used as templates in this chapter contain two 45° angles.

Template

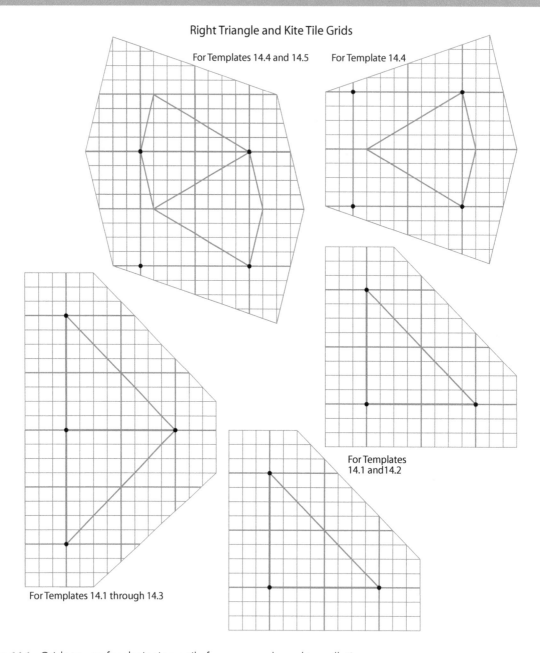

Right Triangle and Kite Tile Grids

For Templates 14.4 and 14.5

For Template 14.4

For Templates 14.1 and 14.2

For Templates 14.1 through 14.3

Figure 14.1. Grids to use for designing a tile for a square-based tessellation.

These can be obtained by simply dividing a square in two along a diagonal. Kites of appropriate dimensions can also fit a square grid.

Kite templates were used by Escher to form some of his most famous designs, including the winged lions (used in the 1946 lithograph *Magic Mirror*) and the riders on horses (used in the 1946 woodcut *Horseman*). The shapes of the kites he used are somewhat different in each design and are not exact fits to a square lattice.

The grids in Figure 14.1 can be used to design tiles using the templates found in this chapter. First choose a template, and then copy the letters and flags for each line segment on one of the grids. Draw curves (these can consist of straight-line segments) connecting the dots in a manner that follows the rules described by the letters and flags. Make sure the curves don't cross anywhere. If you find a tile you like, refine the shape of the curves and add interior details to create a motif for your tessellation. Then cut out your tile as directed for the respective template,

and follow the directions to form a tessellation using the grid in Figure 13.5.

Template 14.1. Right-triangle tessellation with two-fold rotational symmetry

This template, shown in Figure 14.2, has symmetry group *p2* and Heesch type CCC. This tessellation has multiple distinct points of two-fold rotational symmetry.

To draw this tessellation using the grids, cut along three of the six curves making up your tile and cut the other three as straight lines, as shown in Figure 14.2. Form the first set of three lines, as indicated in Figure 14.3, by tracing the three curves with the template fixed in place. Then rotate the template to 180° and form the second set of three lines. Repeat this group of six lines to form a larger tessellation.

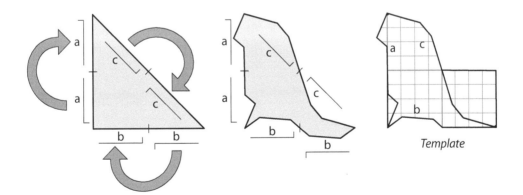

Figure 14.2. Template for a right-triangle-based Escheresque tile that admits a tessellation by two-fold rotations.

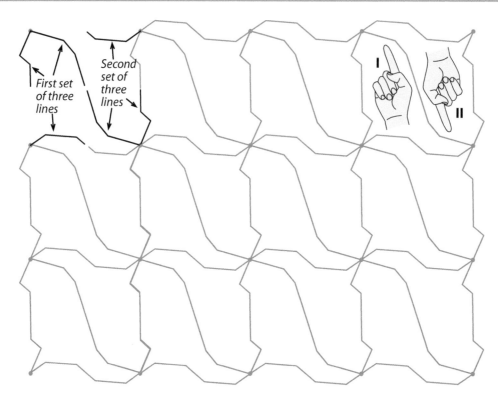

Figure 14.3. Template for drawing a tessellation using a tile designed according to the template of Figure 14.2.

Template 14.2. Right-triangle tessellation with two- and four-fold rotational symmetry

This template, shown in Figure 14.4, has the symmetry group *pgg* and the Heesch type CC_4C_4. Escher's tessellations nos. 35, 118, and 119 are based on this template. The tiles appear in four different orientations in the tessellation. There are two distinct points of four-fold rotational symmetry, with four motifs meeting at one of them and eight at the other.

To draw this tessellation using the grids, cut along the left edge and the top half of the

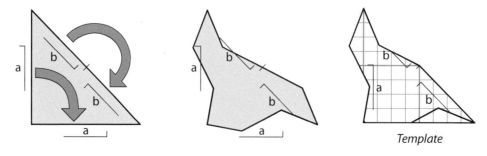

Figure 14.4. Template for a right-triangle-based Escheresque tile that admits a tessellation by two-fold and rotations.

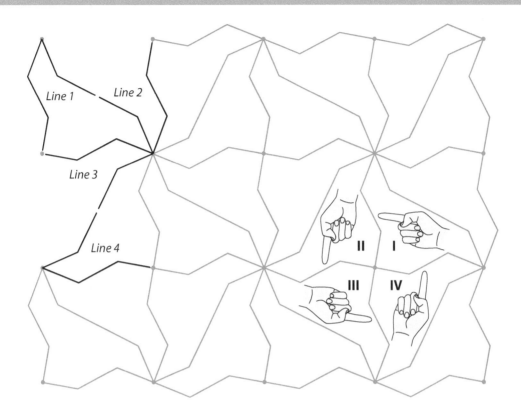

Figure 14.5. Template for drawing a tessellation using a tile designed according to the template of Figure 14.4.

diagonal of your tile, as shown in Figure 14.4. Form Line 1, as indicated in Figure 14.5, by tracing along the cut edges. Then rotate your template to 180° to form Line 2. Next rotate to 90° counter-clockwise to form Line 3. Finally, rotate to 180° to form Line 4. Repeat this set of four lines to form a larger tessellation.

Template 14.3. Right-triangle tessellation with two- and four-fold rotational and reflection symmetry

This template, shown in Figure 14.6, has the symmetry group *p4g*. It has two different tiles, each of which appears in four different orientations. This template can be used to produce one of Escher's most famous tessellations (no. 45), with angels and devils as the motifs.

To draw this tessellation using the grids, completely cut out your top tile, as shown in Figure 14.6. Form Line 1, as indicated in Figure 14.7, by tracing all the way around it. Then rotate it to 90° counter-clockwise to form Line 2. Rotate it to another 90° to form Line 3 and another 90° to form Line 4. Repeat this set of four lines to form a larger tessellation.

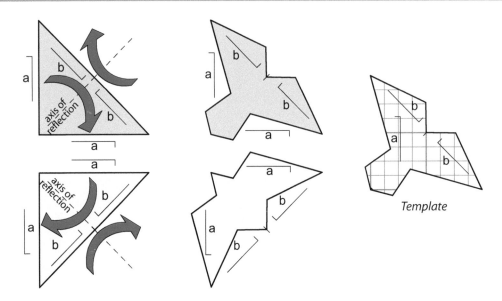

Figure 14.6. Template for a right-triangle-based Escheresque tile that admits a tessellation by two- and four-fold rotation and reflection.

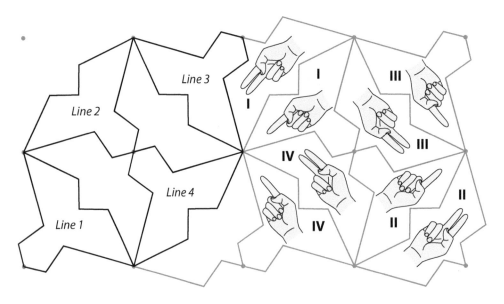

Figure 14.7. Template for drawing a tessellation using a tile designed according to the template of Figure 14.6.

Template 14.4. Kite tessellation with glide reflection symmetry

The particular kite shown in Figure 14.8 can be fit to a square grid. This template has the *pg* symmetry and the Heesch type $G_1G_1G_2G_2$. Escher's 96th numbered tessellation is based on this template. The lengths of the lines a and b can be varied by moving the points P and P′ to the right or left, with the constraint that the distance from P to P′ remain equal to the width of the square. Note that the two

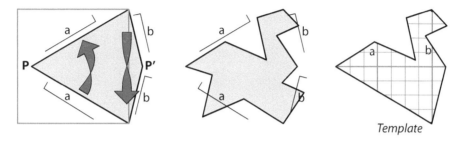

Figure 14.8. Template for a kite-based Escheresque tile that admits a tessellation by glide reflection.

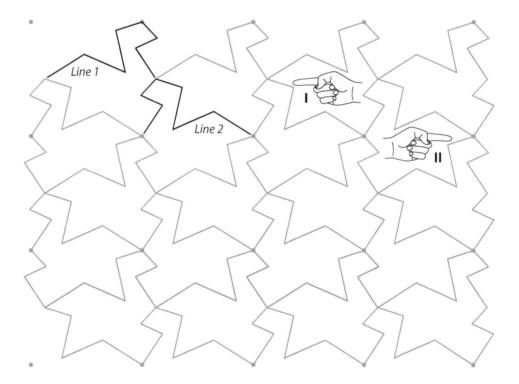

Figure 14.9. Template for drawing a tessellation using a tile designed according to the template of Figure 14.8.

orientations of tiles in the tessellation line up with the grid dots differently.

To draw this tessellation using the grids, cut along the upper edges of your tile, as shown in Figure 14.8. Form Line 1, as indicated on Figure 14.9, by tracing along the cut edges. Then flip your template over from left to right to form Line 2. Repeat this set of two lines to form a larger tessellation.

Template 14.5. Kite tessellation with two motifs and glide reflection symmetry

This template, shown in Figure 14.10, also has the *pg* symmetry, but two different tiles. Each

tile appears in two orientations, mirrored with respect to each other.

To draw this tessellation using the grids, completely cut out your top tile, as shown in Figure 14.10. Form Line 1, as indicated in Figure 14.11, by tracing along the cut edges c and a. Then slide your template over and trace along the two b edges to form Line 2. Next flip your template over from left to right and trace all around it to form Line 3. Repeat this set of three lines to form a larger tessellation.

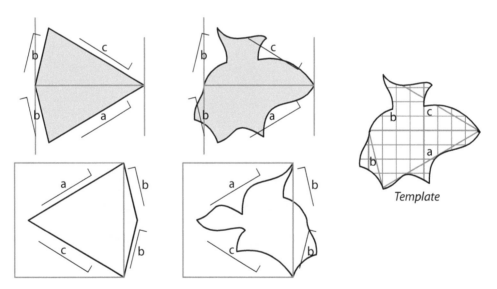

Figure 14.10. Template for a pair of kite-based Escheresque tiles that admit a tessellation by glide reflection.

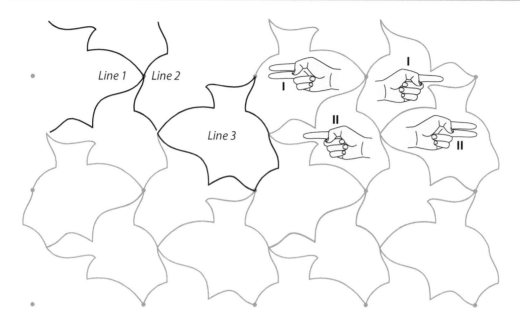

Figure 14.11. Template for drawing a tessellation using a tile designed according to the template of Figure 14.10.

Activity 14.1. Creating a tessellation based on right-triangle tiles

Materials: Copies of template Figures 14.4 and 14.5, Right Triangle and Kite Tile Grids (Figure 14.1) on card stock, Square Tessellation Grid (Figure 13.5) on a paper that can be drawn and colored on, scissors or craft knife, pencil, pen, eraser, and crayons or colored pencils (or other means of coloring a tessellation).

Objective: Learn to create a tessellation by hand based on right-triangle tiles.

Vocabulary: Right triangle, isosceles right triangle.

Activity Sequence

1. Write the vocabulary words on the board and discuss the meaning of each one.
2. Pass out the copies and other materials.
3. On Figure 14.5, have the students draw two different vectors indicating translation distances and directions that would cause the tessellation to perfectly overlap itself.
4. Have the students mark the letters and flags for each of the three right-triangle tile grids (the top half only of the double right-triangle grid) in a similar fashion to how they're marked in Figure 14.4.
5. Using the letters and flags as a guide, have them draw curves connecting the grid dots on at least one of the grids. The two "a" lines should be identical, as should the two "b" lines.
6. Have the students identify a motif in (one of) their shape(s). Then have them refine the shape and rough in interior details.
7. Have them further refine the shape and interior details to produce a prototile.
8. Have them cut out one curve of each type (one for each letter) and cut off the bulk of the sheet to create a template, as shown in Figure 14.4.
9. Have them use their templates to draw the outlines of the tiles on the Square Tessellation Grid sheet, as shown in Figure 14.4, continuing until the sheet is filled up.
10. Have them lightly sketch in key interior details for each tile.
11. Have the students use an ink pen to go over the tile outlines and interior details for the entire tessellation. After the ink is fully dry, unwanted pencil marks can be erased.
12. Have them color their tessellations. Before starting, you may wish to discuss coloring options.

Discussion Questions

1. What motif did you use for your tile? Do you think it was effective? How would you change it if you could do it over again?
2. What step did you find the most challenging? Why?
3. Which tessellation with rotational symmetry do you like better, this one with tiles based on right triangles, or that which was used for Activity 7.2, with tiles based on squares? Why?

Activity 14.2. Creating a tessellation based on kite-shaped tiles

Materials: Copies of template Figures 14.8 and 14.9, Right Triangle and Kite Tile Grids sheet on card stock, Square Tessellation Grid sheet on a paper that can be drawn and colored on, scissors or craft knife, pencil, pen, eraser, and crayons or colored pencils (or other means of coloring a tessellation).

Objective: Learn to create a tessellation by hand based on kite-shaped tiles.

Vocabulary: Kite-shaped tile.

Activity Sequence

1. Write the vocabulary term on the board and discuss its meaning.
2. Pass out the copies and other materials.
3. On Figure 14.9, have the students mark the lines of glide reflection symmetry, along with glide vectors that would cause the tessellation to perfectly overlie itself.
4. Have the students mark the letters and flags for each of the two kite tile grids in a similar fashion to how they're marked on the template sheet.
5. Using the letters and flags as a guide, have them draw curves connecting the grid dots on at least one of the grids. The two "a" lines should be mirror images of each other, as should the two "b" lines.
6. Have the students identify a motif in (one of) their shape(s). Then have them refine the shape and rough in interior details.
7. Have them further refine the shape and interior details to produce a prototile.
8. Have them cut out one curve of each type (one for each letter) and cut off the bulk of the sheet to create a template, as shown in Figure 14.8.
9. Have them use their templates to draw the outlines of the tiles on the Square Tessellation Grid sheet, as directed in Figure 14.8, continuing until the sheet is filled up.
10. Have them lightly sketch in key interior details for each tile.
11. Have the students use an ink pen to go over the tile outlines and interior details for the entire tessellation. After the ink is fully dry, unwanted pencil marks can be erased.
12. Have them color their tessellations. Before starting, you may wish to discuss coloring options.

Discussion Questions

1. What motif did you use for your tile? Do you think it was effective? How would you change it if you could do it over again?
2. What step did you find the most challenging? Why?
3. Which tessellation with glide reflection symmetry do you like better, this one with tiles based on kites, or that which was used for Activity 13.3, with tiles based on squares? Why?

Chapter 15

Escheresque Tessellations Based on Equilateral Triangle Tiles

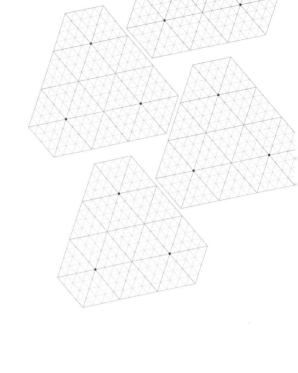

Tessellations based on equilateral triangle tiles are explored in this chapter. Unlike a regular tessellation of squares, a regular tessellation of equilateral triangles contains two different orientations of triangles, related by a 180° rotation. For this reason, there is no single-prototile template that tessellates strictly by translational symmetry.

The fact that equilateral triangles appear in two orientations makes the grid a little more complicated than a square grid. However, tessellations based on equilateral triangles tend to be less static and more interesting than tessellations based on squares. Tiles based on equilateral triangles are particularly well suited for

tessellations with rotational symmetry, and the first template in this chapter has points of two-, three-, and six-fold rotational symmetry.

M.C. Escher used tiles based on equilateral triangles in a number of his tessellations. Most of them have birds and/or fish as the motifs. My personal experience is that distorting equilateral triangle tiles frequently creates tiles that suggest these creatures.

The grids in Figure 15.1 can be used to design tiles using the templates found in this chapter. First, choose a template, and then copy

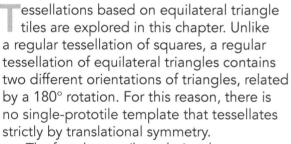

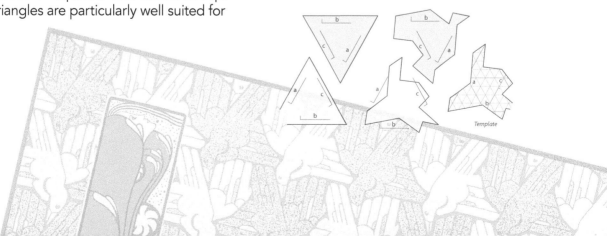

Template

Equilateral Triangle Tile Grids

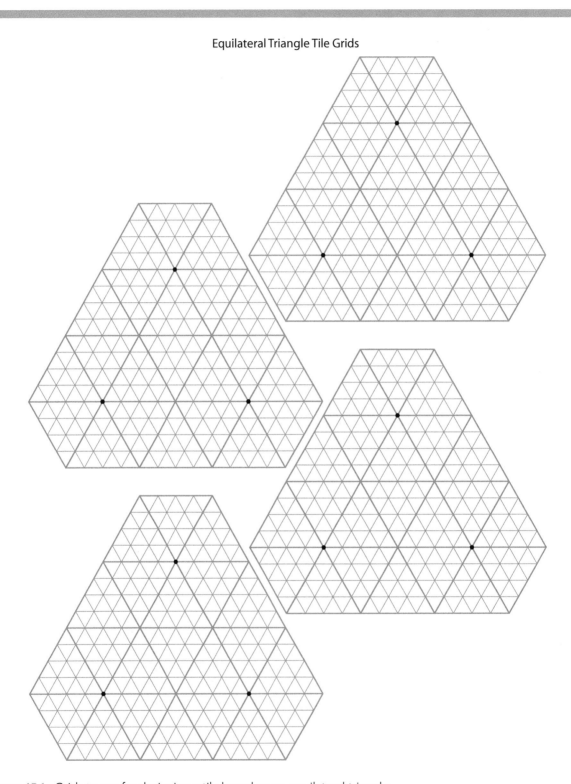

Figure 15.1. Grids to use for designing a tile based on an equilateral triangle.

Equilateral Triangle Tessellation Grid

Figure 15.2. Grid to use for drawing a tessellation based on equilateral triangle tiles.

the letters and flags for each line segment on a grid. Next, draw curves connecting the dots in a manner that follows the rules described by the letters and flags. Then cut out your tile as directed for the chosen template, and follow the directions there to form a tessellation. The grid of Figure 15.2 can be used to draw tessellations based on equilateral triangle tiles.

Template 15.1. Tessellation with six-fold rotational symmetry

This template, shown in Figure 15.3, has the symmetry group *p6* and the Heesch type CC_6C_6. It possesses points of two-, three-, and six-fold rotational symmetry. The tile appears in six different orientations in the tessellation. Escher's tessellations nos. 44, 94, 99, and 100 are based on this template.

To draw this tessellation using the grids, cut along the right edge and right half of the bottom edge of your tile, as shown. Form Line 1, as indicated in Figure 15.4, by tracing along the cut edges. Then rotate your template to 60° (1/6 of a full revolution) counter-clockwise and draw Line 2. Similarly draw Lines 3–6. To form a larger tessellation, repeat this group of six lines. The example below shows where to start your second group by drawing Line 7. Figure 15.5 shows a seahorse design created using this template.

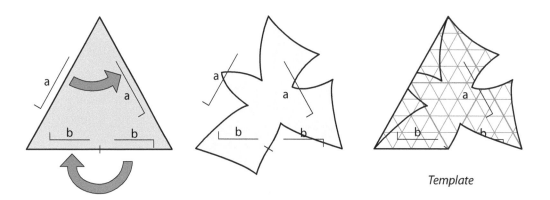

Template

Figure 15.3. Template for an equilateral triangle-based Escheresque tile that admits a tessellation with two-, three-, and six-fold rotations.

Figure 15.4. Template for drawing a tessellation using a tile designed according to the template of Figure 15.3.

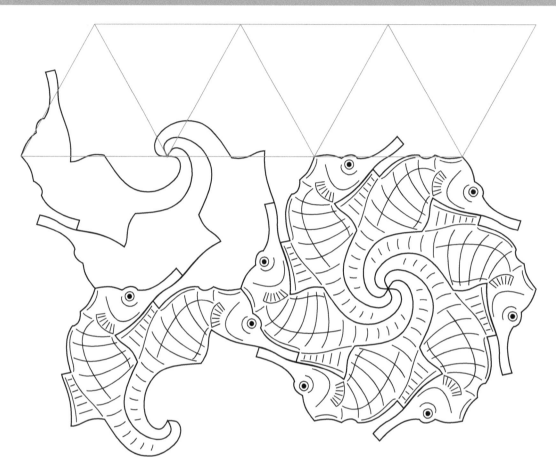

Figure 15.5. This seahorse design is based on Template 15.1. Try completing the seahorses in the top two rows of tiles.

Template 15.2. Tessellation with rotational and glide reflection symmetry

This template, shown in Figure 15.6, has the symmetry group *pgg* and the Heesch type CGG. It has both two-fold rotational symmetry and glide reflection symmetry. The tile appears in four distinct orientations in the tessellations.

To draw this tessellation using the grids, cut along the upper right curve a and the lower right curve b. Trace along these edges to draw Line 1, as shown in Figure 15.7. Then rotate your template to 180° (1/2 of a full revolution) and draw Line 2. Flip your template over from left to right and draw Line 3. Then rotate to 180° and draw Line 4. To form a larger tessellation, repeat this group of four lines as needed.

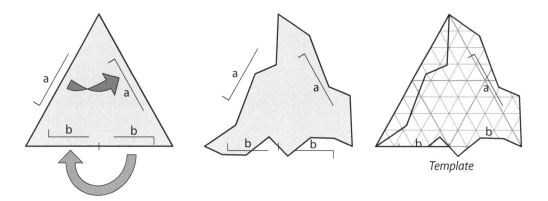

Figure 15.6. Template for an equilateral triangle-based Escheresque tile that admits a tessellation by a combination of rotation and glide reflection.

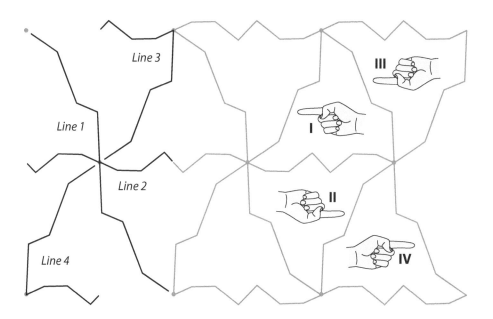

Line 3

Line 1

Line 2

Line 4

I

II

III

IV

Figure 15.7. Template for drawing a tessellation using a tile designed according to the template of Figure 15.6.

Template 15.3. Tessellation with two-fold rotational symmetry only

This template, shown in Figure 15.8, has the symmetry group *p2* and the Heesch type CCC. It has four distinct points of two-fold rotational symmetry. The tile appears in two distinct orientations in the tessellation. Escher's tessellation no. 95 is based on this template.

To draw this tessellation using the grids, cut along three of the six curves making up your tile and cut the other three as straight lines. Form the first set of three lines by tracing the three curves with the template fixed in place, as shown in Figure 15.9. Then rotate the template to 180° and form the second set of three lines. Repeat this group of six lines to form a larger tessellation.

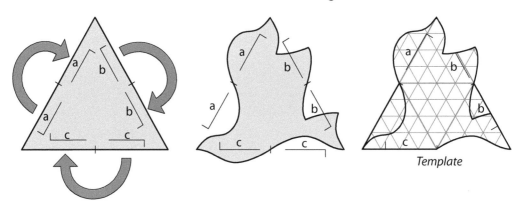

Figure 15.8. Template for an equilateral triangle-based Escheresque tile that admits a tessellation by two-fold rotation.

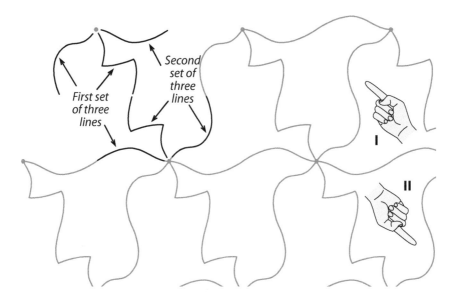

Figure 15.9. Template for drawing a tessellation using a tile designed according to the template of Figure 15.8.

Template 15.4. Tessellation with translational symmetry only

This template, shown in Figure 15.10, has *p1* symmetry. It has two motifs, but only possesses translational symmetry. Escher's tessellations nos. 47, 48, 49, 50, and 52 are based on this template. His motifs in these designs are all birds, fish, and/or frogs.

To draw this tessellation using the grids, fully cut the lower of the two tiles. Trace around it to form a set of three lines, as shown in Figure 15.11. Reposition it as shown to form a second set of three lines. Repeat to form a larger tessellation.

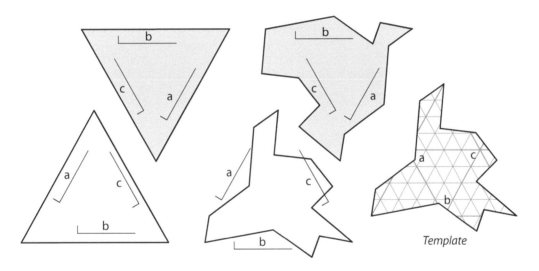

Figure 15.10. Template for a pair of equilateral triangle-based Escheresque tiles that admit a tessellation by translation only.

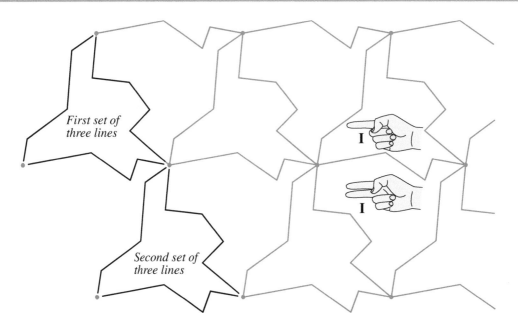

Figure 15.11. Template for drawing a tessellation using a tile designed according to the template of Figure 15.9.

Activity 15.1. Creating an equilateral triangle-based tessellation with rotational symmetry

Materials: Copies of Figures 15.3 and 15.4, copies of the Equilateral Triangle Tile Grids (Figure 15.1) on card stock, copies of the Equilateral Triangle Tessellation Grid (Figure 15.2) on a paper that can be drawn and colored on, scissors or craft knife, pencil, pen, eraser, and crayons or colored pencils (or other means of coloring a tessellation).

Objective: Learn to create a tessellation by hand, based on an equilateral triangle tile, that possesses rotational symmetry.

Vocabulary: Equilateral triangle, *n*-fold rotational symmetry.

Activity Sequence

1. Write the vocabulary terms on the board and discuss their meanings.
2. Pass out the copies and other materials.
3. On Figure 15.4, have the students mark each distinct point of rotational symmetry, using a polygon to indicate the amount of rotation. For two-fold rotation, use a rectangle. For three-fold rotation, use a triangle. For six-fold rotation, use a hexagon.
4. Have the students mark the letters and flags for each of the four Equilateral Triangle Tile Grids in a similar fashion to how they're marked in Figure 15.3.
5. Using the letters and flags as a guide, have them draw curves connecting the grid dots on at least one of the grids. The two "a" lines should be identical, as should the two "b" lines.
6. Have the students identify a motif in (one of) their shape(s). Then have them refine the shape and rough in the interior details.
7. Have them further refine the shape and interior details to produce a prototile.
8. Have them cut out one curve of each type (one for each letter) and cut off the bulk of the sheet to create a template, as shown in Figure 15.3.
9. Have them use their templates to draw the outlines of the tiles on the Equilateral Triangle Tessellation Grid sheet, as directed in Figure 15.4, continuing until the sheet is filled up.
10. Have them lightly sketch in key interior details for each tile.
11. Have the students use an ink pen to go over the tile outlines and interior details for the entire tessellation. After the ink is fully dry, unwanted pencil marks can be erased.
12. Have them color their tessellations. Before starting, you may wish to discuss coloring options.

Discussion Questions

1. What motif did you use for your tile? Do you think it was effective? How would you change it if you could do it over again?
2. What step did you find the most challenging? Why?
3. Did you find designing this tessellation to be more or less difficult than designing a square tessellation with rotational symmetry (Activity 13.2)? Why?
4. Which of the two templates for tessellations possessing rotational symmetry, 13.2 or 15.1, do you like better? Why?

Activity 15.2. Creating an equilateral triangle-based tessellation with glide reflection symmetry

Materials: Copies of Figures 15.6 and 15.7, copies of the Equilateral Triangle Tile Grids (Figure 15.1) on card stock, copies of the Equilateral Triangle Tessellation Grid (Figure 15.2) on a paper that can be drawn and colored on, scissors or craft knife, pencil, pen, eraser, and crayons or colored pencils (or other means of coloring a tessellation).

Objective: Learn to create a tessellation by hand, based on an equilateral triangle tile, that possesses glide reflection symmetry.

Vocabulary: Glide reflection symmetry.

Activity Sequence

1. Write the vocabulary term on the board and discuss its meaning.
2. Pass out the copies and other materials.
3. On Figure 15.7, have the students mark the lines of glide reflection symmetry, along with glide vectors that would cause the tessellation to perfectly overlie itself.
4. Have the students mark the letters and flags for each of the four Equilateral Triangle Tile Grids in a similar fashion to how they're marked in Figure 15.6.
5. Using the letters and flags as a guide, have them draw curves connecting the grid dots on at least one of the grids. The two "b" lines should be identical, while the two "a" lines should be mirror images of each other.
6. Have the students identify a motif in (one of) their shape(s). Then have them refine the shape and rough in the interior details.
7. Have them further refine the shape and interior details to produce a prototile.
8. Have them cut out one curve of each type (one for each letter) and cut off the bulk of the sheet to create a template, as shown in Figure 15.6.
9. Have them use their templates to draw the outlines of the tiles on the Equilateral Triangle Tessellation Grid sheet, as directed on the template of Figure 15.2, continuing until the sheet is filled up.
10. Have them lightly sketch in key interior details for each tile.
11. Have the students use an ink pen to go over the tile outlines and interior details for the entire tessellation. After the ink is fully dry, unwanted pencil marks can be erased.
12. Have them color their tessellations. Before starting, you may wish to discuss coloring options.

Discussion Questions

1. What motif did you use for your tile? Do you think it was effective? How would you change it if you could do it over again?
2. What step did you find the most challenging? Why?
3. Did you find designing this tessellation to be more or less difficult than designing the tessellation in Activities 15.1? Why?
4. Which of the two templates for tessellations possessing glide reflection symmetry, 13.3 or 15.2, do you like better? Why?

Chapter 16

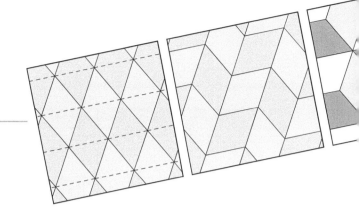

Escheresque Tessellations Based on 60°–120° Rhombus Tiles

In this chapter, some tessellations based on 60°–120° rhombus tiles are described. These rhombi are made up of two equilateral triangles, as shown on the left in Figure 16.1, and can be drawn on the Equilateral Triangle Tessellation Grid discussed in Chapter 15. All rhombi can be oriented the same, so tessellations possessing only translational symmetry are allowed. There are other ways of arranging the same rhombi, though. One possibility, which is used for Template 16.4, is shown in the center in Figure 16.1. A three-color possibility, which is

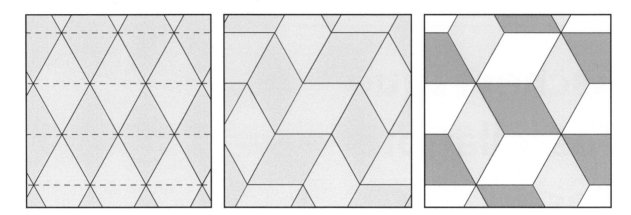

Figure 16.1. Three different geometric tessellations formed from 60°–120° rhombi. Note each rhombus can be divided into two equilateral triangles.

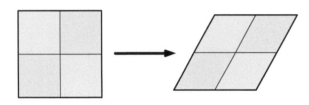

Figure 16.2. A 60°–120° rhombus can be created by skewing a square.

used for Templates 16.3 and 16.6, is shown in the right in Figure 16.1.

A rhombus has four sides of equal length and two different interior angles. A square can be considered a special case of a rhombus, in which all four interior angles are equal.

Because they are closely related, some of the same Heesch types can be used as in Chapter 13. For example, Template 16.1 is the same Heesch type as Template 13.1 and can be considered a skewed version of Template 13.1 (Figure 16.2).

Use these grids in Figure 16.3 to design tiles using the templates found in this chapter. First choose a template, and then copy the letters and flags for each line segment on a grid. Next, draw curves connecting the dots in a manner that follows the rules described by the letters and flags. Then cut out your tile as directed for the chosen template, and follow the directions to form a tessellation.

60° - 120° Rhombus Tile Grids

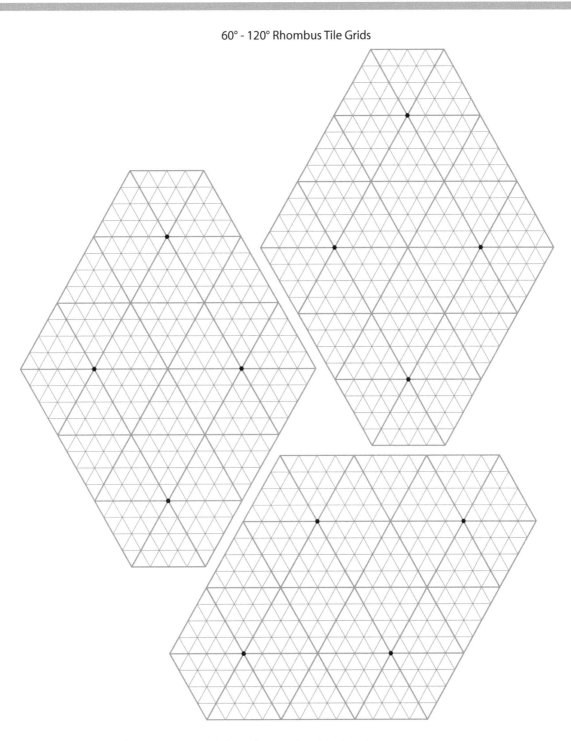

Figure 16.3. Grids to use for designing a tile based on a 60°–120° rhombus.

Template 16.1. Tessellation with translational symmetry only

This template has the symmetry group *p*1 and the Heesch type TTTT. This is the simplest way to form a tessellation using an equilateral triangle grid. There are two independent line segments, each of which simply translates from one side of the rhombus to the other, as shown in Figure 16.4.

To draw this tessellation, cut along the top and right edges of your tile to form your template. Form Line 1 by tracing the cut edges, as shown in Figure 16.5. Repeatedly form this same line in the same orientation to form a larger tessellation.

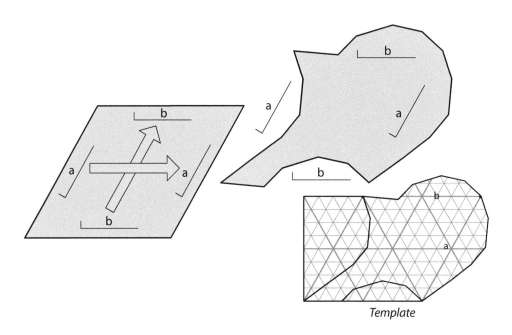

Template

Figure 16.4. Template for a rhombus-based Escheresque tile that admits a tessellation by translation only.

Line 1

Figure 16.5. Template for drawing a tessellation using a tile designed according to the template of Figure 16.4.

Template 16.2. Tessellation with reflection symmetry

This template, shown in Figure 16.6, has the symmetry group *pm*. It uses a single line segment four times to create a tile and a tessellation with reflection symmetry.

To draw this tessellation using the grids, cut along the upper right curve of your tile to form your template. Form Line 1, as indicated in Figure 16.7, by tracing the cut curve. Then flip your template over from left to right to form Line 2. Repeat as needed to form a larger tessellation.

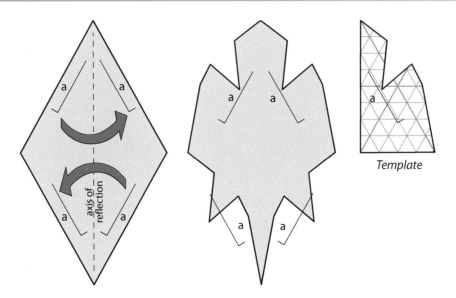

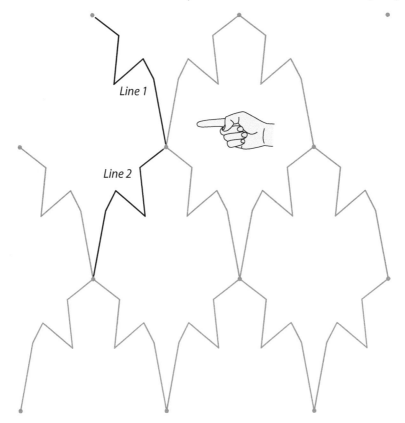

Figure 16.6. Template for a rhombus-based Escheresque tile that admits a tessellation by simple translation.

Figure 16.7. Template for drawing a tessellation using a tile designed according to the template of Figure 16.6.

Template 16.3. Tessellation with three-fold rotational symmetry

This template, shown in Figure 16.8, has the symmetry group *p3* and the Heesch type $C_3C_3C_3C_3$. It forms a tessellation with three distinct points of three-fold rotational symmetry. Escher's tessellations nos. 4 and 21 are based on this template.

To draw this tessellation, cut along the upper left and upper right edges of your tile to form your template. Form Line 1 as indicated in Figure 16.9 by tracing the cut edges. Rotate your template to 120° counter-clockwise to draw Line 2. Rotate again to form Line 3. Repeatedly form this same group of three lines to form a larger tessellation.

This tessellation of Figure 16.10 is based on Template 16.3. Notice how a line is used to move a portion of each tile away from a vertex. Try completing the drawing and coloring of the tessellation.

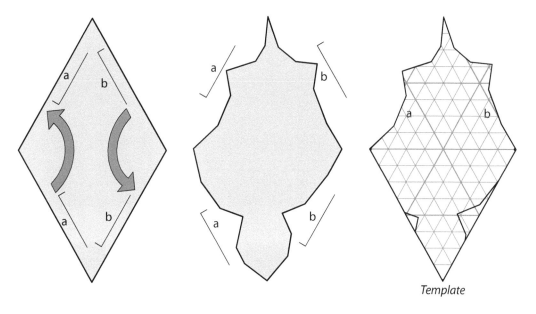

Figure 16.8. Template for a rhombus-based Escheresque tile that admits a tessellation by rotation.

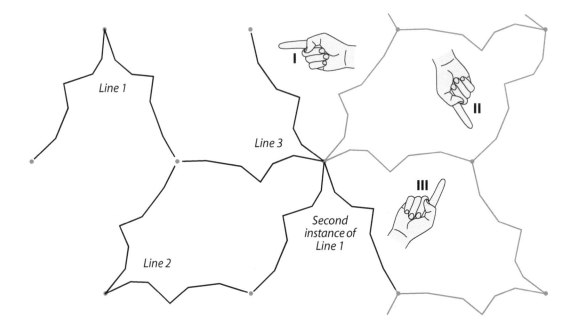

Figure 16.9. Template for drawing a tessellation using a tile designed according to the template of Figure 16.8.

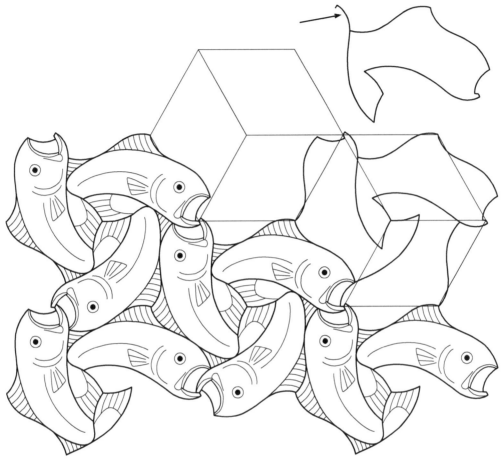

Figure 16.10. A tessellation of fish based on Template 16.3.

Template 16.4. Tessellation with glide reflection symmetry

This template, shown in Figure 16.11, has the symmetry group *pg* and the Heesch type TGTG. It has glide reflection symmetry, and the tile appears in two distinct orientations in the tessellation.

To draw this tessellation, cut along the upper left and upper right curves to form your template. Trace the cut edges to form Line 1, as shown in Figure 16.12. Then flip your template from top to bottom, rotate to 60° clockwise, and trace Line 2. Repeat these steps to form a larger tessellation.

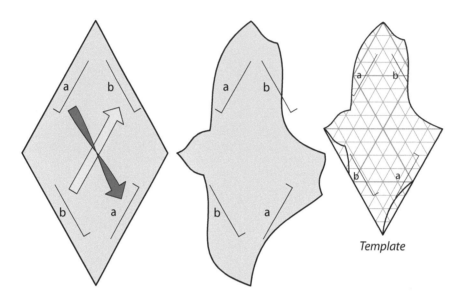

Figure 16.11. Template for a rhombus-based Escheresque tile that admits a tessellation by glide reflection.

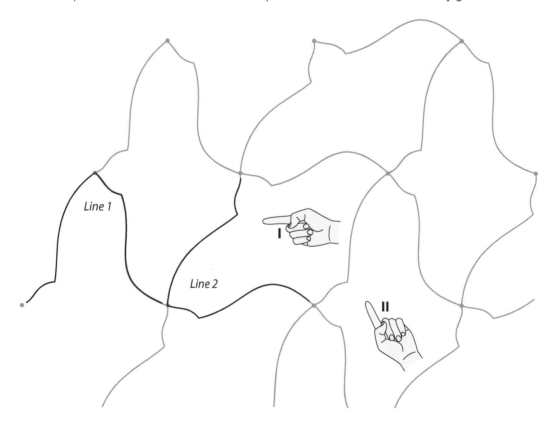

Figure 16.12. Template for drawing a tessellation using a tile designed according to the template of Figure 16.11.

Template 16.5. Rhombus tessellation with rotational and glide reflection symmetry

This template has the *cmm* symmetry. It uses a single line segment eight times to create two different tiles and a tessellation

with reflection symmetry, as shown in Figure 16.13.

To draw this tessellation, fully cut the lower part of your two tiles. Trace around it to form a set of four lines, as shown in Figure 16.14. Flip your tile over from top to bottom, and trace around it to form a second set of four lines. Repeat as needed to form a larger tessellation.

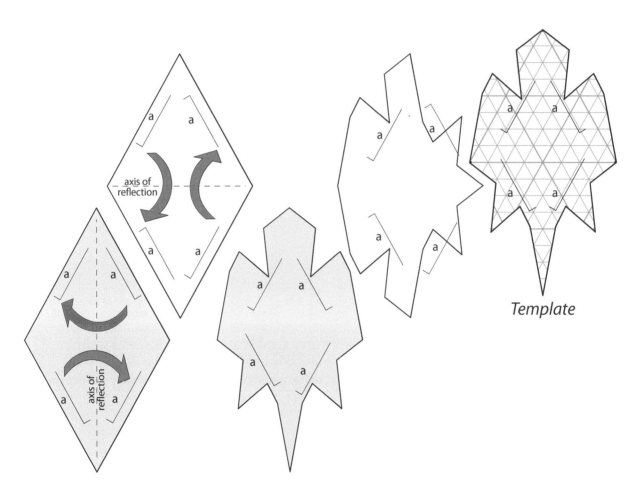

Template

Figure 16.13. Template for a pair of rhombus-based Escheresque tiles that admit a tessellation by translation only.

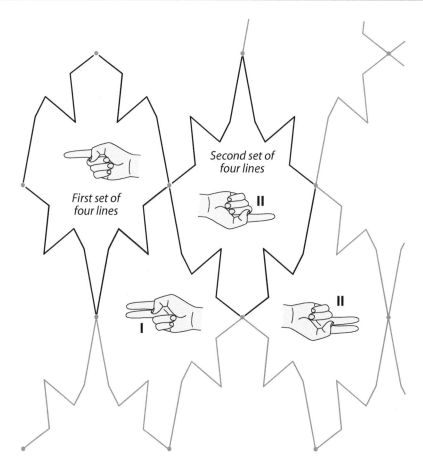

Figure 16.14. Template for drawing a tessellation using a tile designed according to the template of Figure 16.13.

Template 16.6. Tessellation with kaleidoscopic symmetry

This template has *p3m1* symmetry. It creates a tessellation with three different motifs and multiple lines of reflection symmetry, as shown in Figure 16.15. Escher's tessellations nos. 69, 85, 103, and 123 are based on this template. This sort of tessellation could be produced by a kaleidoscope with three mirrors arranged as an equilateral triangle, which is the most common type of kaleidoscope.

To draw a tessellation using this template, make two templates, as shown in Figure 16.15. Note that the Equilateral Triangle Tile Grids of Chapter 15 actually work better for this tessellation than the tile grids in this chapter. Trace completely around the top template to form Line 1, as shown in Figure 16.16. Rotate your template to 120° clockwise to form Line 2, and again to form Line 3. Using the lower template, trace around the two cut "c" edges to form Line 4. Rotate your template to 120° clockwise to form Line 5, and rotate again to form Line 6. Repeat as needed to form a larger tessellation.

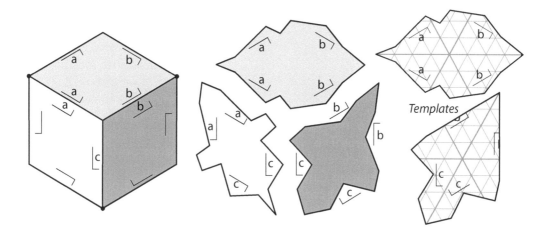

Figure 16.15. Template for a trio of rhombus-based Escheresque tiles that admit a tessellation by reflection.

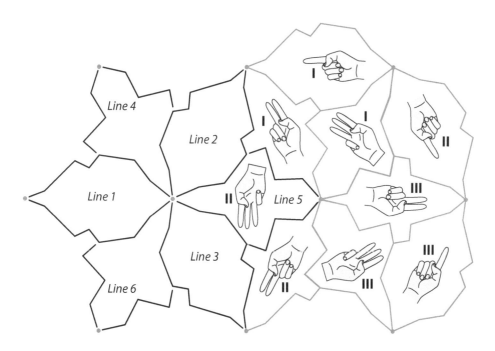

Figure 16.16. Template for drawing a tessellation using a tile designed according to the template of Figure 16.14.

Activity 16.1. Creating a tessellation with bilaterally symmetric tiles

Materials: Copies of Figures 16.6 and 16.7, copies of the 60°–120° Rhombus Tile Grids (Figure 16.3) on card stock, copies of Equilateral Triangle Tessellation Grid sheet (Figure 15.2) on a paper that can be drawn and colored on, scissors or craft knife, pencil, pen, eraser, and crayons or colored pencils (or other means of coloring a tessellation).

Objective: Learn to create a tessellation by hand, based on a 60°–120° rhombus, that has bilaterally symmetric tiles.

Vocabulary: Rhombus, bilateral symmetry.

Activity Sequence

1. Write the vocabulary terms on the board and discuss their meanings.
2. Pass out the copies and other materials.
3. On Figure 16.7, have the students mark the lines of reflection symmetry and lines of glide reflection symmetry, along with glide vectors that would cause the tessellation to perfectly overlie itself.
4. Have the students mark the letters and flags for each of the three Rhombus Tile Grids in a similar fashion to how they're marked in Figure 16.6.
5. Using the letters and flags as a guide, have them draw curves connecting the grid dots on at least one of the grids. Two of the four lines should be identical, and two should be mirror images of the other lines.
6. Have the students identify a motif in (one of) their shape(s). Then have them refine the shape and rough in the interior details.
7. Have them further refine the shape and interior details to produce a prototile.
8. Have them cut out one curve and cut off the bulk of the sheet to create a template, as shown in Figure 16.6.
9. Have them use their templates to draw the outlines of the tiles on the Equilateral Triangle Tessellation Grid sheet, as directed in Figure 16.7, continuing until the sheet is filled up.
10. Have them lightly sketch in key interior details for each tile.
11. Have the students use an ink pen to go over the tile outlines and interior details for the entire tessellation. After the ink is fully dry, unwanted pencil marks can be erased.
12. Have them color their tessellations. Before starting, you may wish to discuss coloring options.

Discussion Questions

1. What motif did you use for your tile? Do you think it was effective? How would you change it if you could do it over again?
2. Did you find designing this tessellation to be more or less difficult than designing the tessellations in earlier activities, in which the tiles did not possess bilateral symmetry? Why?

Activity 16.2.
Creating a tessellation with kaleidoscopic symmetry

Materials: Copies of Figures 16.15 and 16.16, Equilateral Triangle Tile Grids sheet (Figure 15.1) on card stock, Equilateral Triangle Tessellation Grid sheet (Figure 15.2) on a paper that can be drawn and colored on, scissors or craft knife, pencil, pen, eraser, and crayons or colored pencils (or other means of coloring a tessellation).

Objective: Learn to create a tessellation by hand, based on a 60°–120° rhombus, that has three different tiles and kaleidoscope symmetry.

Vocabulary: Kaleidoscope symmetry.

Activity Sequence

1. Write the vocabulary term on the board and discuss its meaning. If possible, have the students look through a kaleidoscope to better understand the symmetry of this tessellation.
2. Pass out the copies and other materials.
3. On Figure 16.16, have the students mark the lines of reflection symmetry and points of rotational symmetry with equilateral triangles.
4. Have the students mark the letters and flags for each of the four Equilateral Triangle Tile Grids in a similar fashion to how they're marked in Figure 16.15. Turning the grid sheet upside down will orient the tile grids the same as the tiles in Figure 16.15. Each tile grid will contain a set of

three different prototiles that appear in the tessellation.

5. Using the letters and flags as a guide, have them draw curves to create a set of three tiles on at least one of the grids. Note that there are three distinct lines, each of which appears both in reflected and unreflected orientations.
6. Have the students identify motifs in one set of three shapes. Then have them refine the shapes and rough in the interior details.
7. Have them further refine the shapes and interior details to produce a set of three prototiles.
8. Have them cut out the templates, as directed in Figure 16.15.
9. Have them use their templates to draw the outlines of the tiles on the Equilateral Triangle Tessellation Grid sheet, as directed in Figure 16.16, continuing until the sheet is filled up.
10. Have them lightly sketch in key interior details for each tile.
11. Have the students use an ink pen to go over the tile outlines and interior details for the entire tessellation. After the ink is fully dry, unwanted pencil marks can be erased.
12. Have them color their tessellations. Before starting, you may wish to discuss coloring options.

Discussion Questions

1. What motifs did you use for your tiles? Do the three motifs go together somehow? How would you change it if you could do it over again?
2. What sort of motifs do you think work well for a tessellation with kaleidoscope symmetry? Why?

Chapter 17

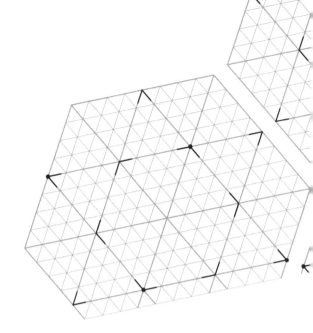

Escheresque Tessellations Based on Hexagonal Tiles

In this chapter, some tessellations based on hexagon and hexagram tiles are described. These tessellations can be drawn on the Equilateral Triangle Tessellation Grid in Chapter 15. A wide variety of tessellations may be formed using regular hexagons and hexagrams along with equilateral triangles. Some examples of tessellations incorporating regular hexagons and hexagrams are shown in Chapter 2.

The grids in Figure 17.1 should be used to design tiles using the templates found in this chapter. First choose a template, and then copy the letters and flags for each line segment on a grid. The short black lines show how the corners of hexagons and hexagrams fit on the grid. Draw curves connecting the corners of the tiles in a manner that follows the rules described by the letters and flags. Then cut out your tile as directed for the chosen template, and follow the directions to form a tessellation.

Template

Hexagon Tile Grids

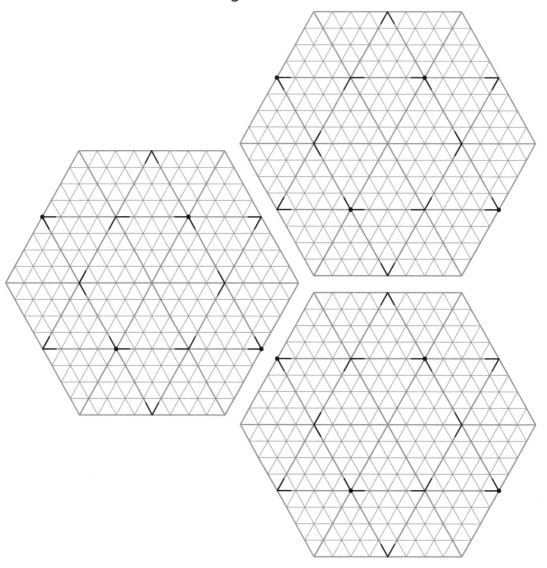

Figure 17.1. Grids for designing tiles based on hexagons and hexagrams.

Template 17.1. Tessellation with three-fold rotational symmetry

This template has the symmetry group *p6* and the Heesch type $C_3C_3C_3C_3C_3C_3$. It has three distinct points of three-fold rotational symmetry, and the tile appears in three different orientations. Escher's famous lizard tessellation, no. 25, is based on this template.

To draw this tessellation, cut along three of the six curves making up your tile and cut the other three as straight lines, as shown in Figure 17.2. Form the first set of three lines by tracing the three curves with your template fixed in place, as shown in Figure 17.3. Then rotate the template to 120° counter-clockwise and form the second set of three lines, as shown in the figure, and another 120° to form the third set. Repeat this group of nine lines to form a larger tessellation.

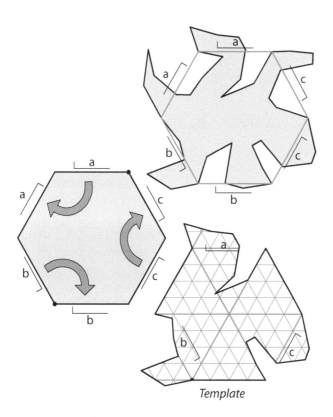

Template

Figure 17.2. Template for a hexagon-based Escheresque tile that admits a tessellation by rotation.

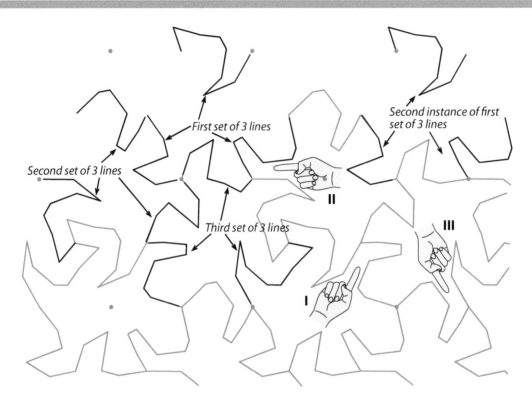

Figure 17.3. Template for drawing a tessellation using a tile designed according to the template of Figure 17.1.

Template 17.2. Tessellation with six-fold rotational symmetry

This template, shown in Figure 17.4, has the symmetry group *p6*. The tiles in this tessellation have six-fold rotational symmetry. This sort of symmetry works well for snowflakes, starfish, flowers, and some other plants. A regular hexagon can be thought of as a group of six equilateral triangles. Using the triangles in the Equilateral Triangle Tessellation Grid would result in hexagons that are too big for tessellating on a single sheet of paper, and so the hexagon here is half scale. For this reason, only two of the six corners will lie on the black dots on the grid. This tessellation, with a prototile in which the same line segment appears 12 times, has points of two-, three-, and six-fold rotational symmetry.

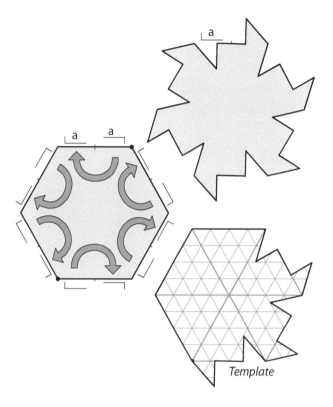

Figure 17.4. Template for a six-fold hexagon-based Escheresque tile that admits a tessellation by rotation.

To draw a tessellation, make a template by cutting out half of a full tile, as shown in Figure 17.4. Draw Line 1 on the grid by tracing along the cut edges of your template, as shown in Figure 17.5. Translate your template to draw Lines 2–4. Note that each of the four lines matches up to the dots on the grid differently. Repeat this group of four lines as needed to form a larger tessellation.

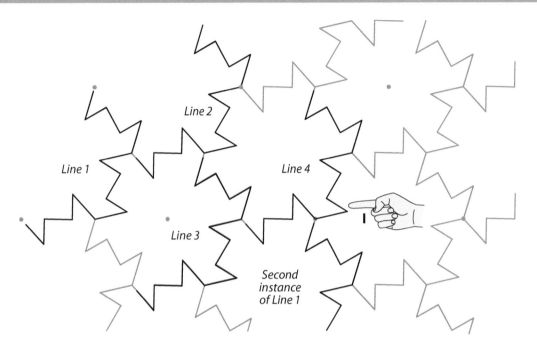

Figure 17.5. Template for drawing a tessellation using a tile designed according to the template of Figure 17.4.

Template 17.3. Tessellation based on hexagons and hexagrams

This template, shown in Figure 17.6, has the symmetry group *p6*. It also works well for objects with six-fold symmetry, but it has two prototiles, compared to the single prototile of Template 17.2. One is a regular hexagon, and the other is a hexagram. The left and middle portions in Figure 17.6 are reduced in size to fit better on the page, but the template is of full size.

To draw this tessellation, completely cut out the smaller tile to form your template, as shown in Figure 17.6. Trace all the way

Figure 17.6. Template for a pair of tiles based on a hexagon and hexagram that tile with one another.

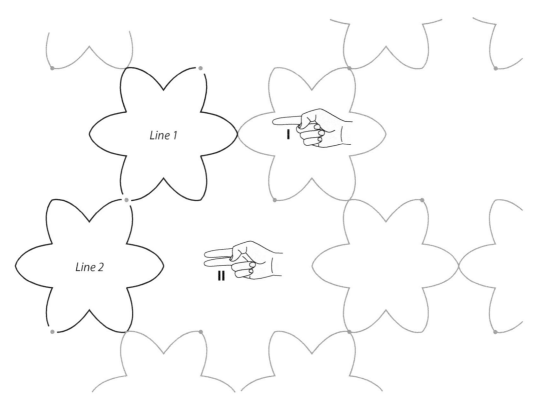

Line 1

Line 2

I

II

Figure 17.7. Template for drawing a tessellation using a tile designed according to the template of Figure 17.6.

around your template to form Line 1, as shown in Figure 17.7. Translate your template to draw Line 2. Repeat this group of two lines as needed to form a larger tessellation.

The snowflake tessellation of Figure 17.8 is based on Template 17.3. Try completing the drawing with your own snowflake structure and then color the tessellation.

Figure 17.8. A tessellation of snowflakes based on Template 17.3.

Activity 17.1. Creating a tessellation based on hexagonal tiles

Materials: Copies of Figures 17.2 and 17.3, Hexagon Tile Grids (Figure 17.1) on card stock, Equilateral Triangle Tessellation Grid sheet (Figure 15.2) on a paper that can be drawn and colored on, scissors or craft knife, pencil, pen, eraser, and crayons or colored pencils (or other means of coloring a tessellation).

Objective: Learn to create a tessellation by hand, based on a hexagonal tile, that has three-fold rotational symmetry.

Vocabulary: Hexagonal tile, corners of a tile.

Activity Sequence

1. Write the vocabulary terms on the board and discuss their meanings.
2. Pass out the copies and other materials.
3. On Figure 17.3, have the students mark each distinct point of three-fold rotational symmetry with a triangle.
4. Have the students mark the letters and flags for each of the three Hexagon Tile Grids in a similar fashion to how they're marked on the template sheet.
5. Using the letters and flags as a guide, have them draw curves connecting the corners of the tiles on at least one of the grids. The two "a" lines should be identical, as should the two "b" lines and "c" lines.
6. Have the students identify a motif in (one of) their shape(s). Then have them refine the shape and rough in the interior details.
7. Have them further refine the shape and interior details to produce a prototile.
8. Have them cut out one curve of each type (one for each letter) and cut off the bulk of the sheet to create a template, as shown in Figure 17.2.
9. Have them use their templates to draw the outlines of the tiles on the Equilateral Triangle Tessellation Grid sheet, as directed in Figure 17.3, continuing until the sheet is filled up.
10. Have them lightly sketch in key interior details for each tile.
11. Have the students use an ink pen to go over the tile outlines and interior details for the entire tessellation. After the ink is fully dry, unwanted pencil marks can be erased.
12. Have them color their tessellations. Before starting, you may wish to discuss coloring options.

Discussion Questions

1. What motif did you use for your tile? Do you think it was effective? How would you change it if you could do it over again?
2. What step did you find the most challenging? Why?
3. Which of the three templates for tessellations possessing rotational symmetry, 13.2, 15.1, or 17.1, do you like the best? Why?

Activity 17.2. Creating a hexagon-based tessellation with two-, three-, and six-fold rotational symmetry

Materials: Copies of Figures 17.4 and 17.5, Hexagon Tile Grids (Figure 17.1) on card stock, Equilateral Triangle Tessellation Grid sheet on a paper that can be drawn and colored on, scissors or craft knife, pencil, pen, eraser, and crayons or colored pencils (or other means of coloring a tessellation).

Objective: Learn to create a tessellation by hand, based on a hexagonal tile, that has two-, three-, and six-fold rotational symmetry.

Vocabulary: Hexagonal tile, *n*-fold rotational symmetry, corners of a tile.

Activity Sequence

1. Write the vocabulary terms on the board and discuss their meanings.
2. Pass out the copies and other materials.
3. On Figure 17.5, have the students mark each point of two-fold rotational symmetry with a rectangle, each point of three-fold rotational symmetry with a triangle, and each point of six-fold rotational symmetry with a hexagon.
4. Have the students mark the letters and flags for each of the three Hexagon Tile Grids in a similar fashion to how they're marked on the template sheet. Also have them make tick marks halfway along each edge of the hexagon.
5. Using the letters, flags, and tick marks as a guide, have them draw curves connecting the corners of the tiles on at least one of the grids. Each edge of the hexagon should have two short "a" lines that are related by a 180° rotation about a tick mark.
6. Have the students identify a motif in (one of) their shape(s). Then have them refine the shape and rough in the interior details.
7. Have them further refine the shape and interior details to produce a prototile.
8. Have them cut out half of a full tile to create a template, as shown in Figure 17.4.
9. Have them use their templates to draw the outlines of the tiles on the Equilateral Triangle Tessellation Grid sheet, as directed in Figure 17.5, continuing until the sheet is filled up.
10. Have them lightly sketch in key interior details for each tile.
11. Have the students use an ink pen to go over the tile outlines and interior details for the entire tessellation. After the ink is fully dry, unwanted pencil marks can be erased.
12. Have them color their tessellations. Before starting, you may wish to discuss coloring options.

Discussion Questions

1. What motif did you use for your tile? Do you think it was effective? How would you change it if you could do it over again?
2. What step did you find the most challenging? Why?
3. In Activity 15.1, a different tessellation with two-, three-, and six-fold symmetry was created. Which of the two do you like better? Why?

Chapter 18

Decorating Tiles to Create Knots and Other Designs

A **decorated tile** *is one which has mark-ings in its interior.* The features in Escheresque tiles such as eyes that make them look more lifelike are a form of deco-ration. The markings on Penrose tiles that enforce matching rules are another exam-ple. In this chapter, the use of decoration to create other types of designs, including knots and links, will be described.

The role of combinatorics

Combinatorics is the branch of mathematics that deals with enumerating combinations, including the number of ways a particular set of tiles can fit together. In the context of til-ing, this topic traces back to a 1704 memoir by Sébastien Truchet that describes what are

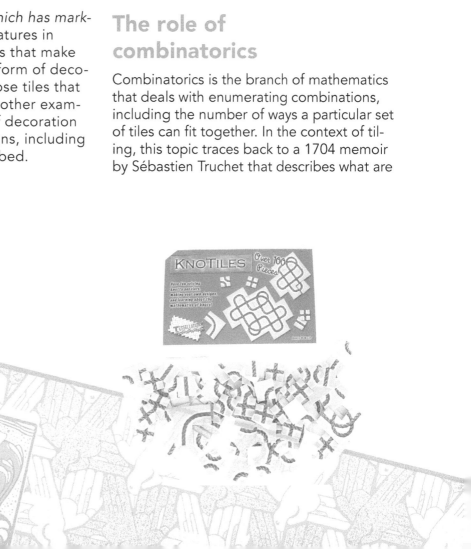

now known as Truchet tiles [Wikipedia 2019]. In the form described by Truchet, a square tile is divided along its diagonal into black and white halves (Figure 18.1). A popular variation using quarter circles is also shown in Figure 18.1.

Note there are four distinct orientations for the left tile in Figure 18.1 but only two for the right tile. For a 2 × 2 patch of four tiles, there are four or two (depending which tile is used) choices for the orientation of the first tile, for each of those first tiles four or two choices for the second tile, etc. This results in $4^4 = 256$ possibilities for a patch of four of the original Truchet tiles, but only $2^4 = 16$ possibilities for the tile with quarter circles. It turns out many of these patches are identical within an overall rotation of the patch. For example, with the quarter-circle tiles, only 6 of the 16 are distinct.

M.C. Escher played around with these ideas with a set of "ribbon blocks" [Schattscheider 2004]. Escher made a square printing block with "ribbons" crossing on it. The design was not mirror symmetric, and there were two possibilities for an over-under crossing of ribbons. This resulted in four distinct blocks, each of which could be printed

in four orientations on a square grid. These can be explored with an online application [Passsiouras 2019].

In 1995, I used these ideas to design a puzzle called "21 Faces". The puzzle has identical square tiles decorated with facial features (Figure 18.2a). A face seen in profile is formed surrounding a vertex. There are two simple rules for using the tiles: (1) each face must contain only one eye and (2) a face that forces two eyes on an adjacent face is not allowed. Under these rules, 21 distinct faces can be formed. A wide variety of tessellations

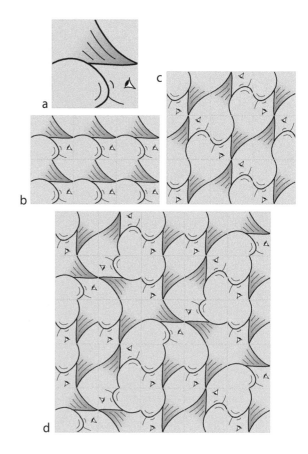

Figure 18.2. A single square tile decorated with facial features can form a wide variety of tessellations of faces.

Figure 18.1. The original Truchet tiles (left) and a more recent variation, with examples of the sort of pattern formed by patches of the tiles.

are possible using the pieces. In Figure 18.2b, a single face tessellates by translation. In Figure 18.2c a single face is formed, but it appears in two orientations. In Figure 18.2d, a group of six different faces tessellates by translation.

Using tessellations to create knots and links

Like tessellations, knots have utilitarian, decorative, and mathematics applications. Knots are used for tying shoelaces, rigging sails on ships, and countless other practical tasks. They play a central role in Celtic and Islamic art, as well as in knitting and other fiber arts. From a mathematics perspective, *a **knot** is a closed curve in a three-dimensional Euclidean space that doesn't intersect itself*. The mathematical field of knot theory is closely related to graph theory.

Figure 18.3. The canonical projections of the trefoil and figure-eight knots.

Knots are classified according to their number of crossings. The simplest possible knot has three crossings and is known as the trefoil. The next simplest knot has four crossings and is known as the figure-eight (Figure 18.3). A two-dimensional drawing of a knot like in Figure 18.3 is considered a projection of a three-dimensional knot. The number of distinct knots increases rapidly with the number of crossings, from one type for three or four crossings to two types for five crossings, three types for six crossings, seven types for seven crossings, etc. *Two or more tangled knots are called a **link***; i.e., a knot has to consist of a single strand. Knots are usually depicted with an alternating over/under weave.

Any knot can be depicted in two dimensions using decorated tiles. Various people have explored this idea. Lomonaco and Kauffman [Lomonaco and Kauffman 2008] developed a "quantum knot system" for treating knots mathematically. Their system uses the five square tiles shown in Figure 18.4. In 2007, I independently designed a puzzle called "KnoTiles" based on the same set of tiles, but omitting the blank tile and adding a couple of edge tiles that allow more aesthetically pleasing designs (Figure 18.5). Any knot can be depicted using a patch of tiles of the type shown in Figure 18.4. For example,

Figure 18.4. Five square tiles that can be used to construct a projection of any knot.

Figure 18.5. KnoTiles puzzle.

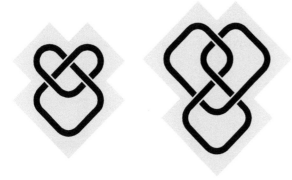

Figure 18.6. The trefoil and figure-eight knots constructed using square knot tiles.

square-tile versions of the knots of Figure 18.3 are shown in Figure 18.6.

Any tile can be decorated with knot graphics, of course. Square tiles are the simplest to handle mathematically, but other shapes can yield more attractive designs. A puzzle called "Tantrix", based on hexagonal tiles with knot graphics, appeared around 1990 [Tantrix 2020]. In Figure 18.7, square knots were placed in each tile of a Cairo pentagon tiling. This results in an infinite periodic pattern of four-lobed strands, which can be thought of as a link with an infinite number of strands.

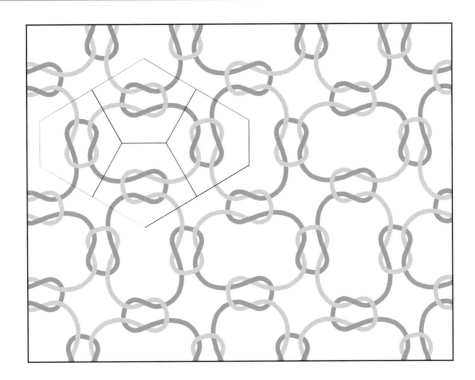

Figure 18.7. A link design created by decorating each pentagon in a Cairo pentagon tiling with knot graphics. The edges of four of the tiles are shown in black.

Creating iterated and fractal knots and links with fractal tilings

The tiles in fractal tilings can also be decorated with knot graphics. The simplest type of fractal tiling has only one singular point, such as the tiling of Figure 18.8a, with four generations of the tile shown. Combining a pair of these tiles gives the tile of Figure 18.8b, which has been decorated with knot graphics. There are two versions of the tile, with the crossings inverted. This allows the strands to alternate in an over/under weave, as shown in Figure 18.8c. Making a finite knot from these four rings of tiles is possible by connecting the loose ends in the center and at the edges, as shown in Figure 18.9.

For a typical fractal tiling, the prototile has two edge lengths, since tiles must mate to tiles of the next size larger and smaller. For a strand to be smoothly continuous across the border between tiles of different sizes, the strand width needs to vary across the tile, as seen in Figure 18.8b.

A mathematical fractal has infinite levels of detail. The knot of Figure 18.9 is finite, but it was created by the sort of iterative process used to generate fractals. A knot of this sort is sometimes referred to as an iterated knot.

A related type of graphical decoration is rings. In Figure 18.10, ring graphics have been applied to a kite-shaped tile, resulting in a fractal arrangement of rings, a fractal link. Note that each tile has rings of two different sizes. The colors are reversed between successive generations so that each ring is

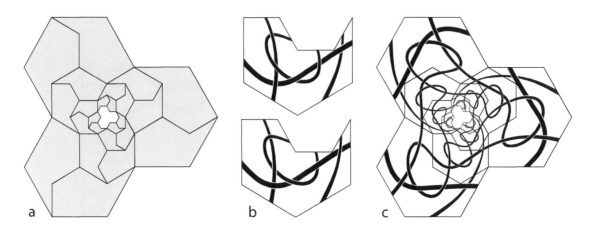

a   b   c

Figure 18.8. Decorating a tile that admits a tiling with a singular point at the center.

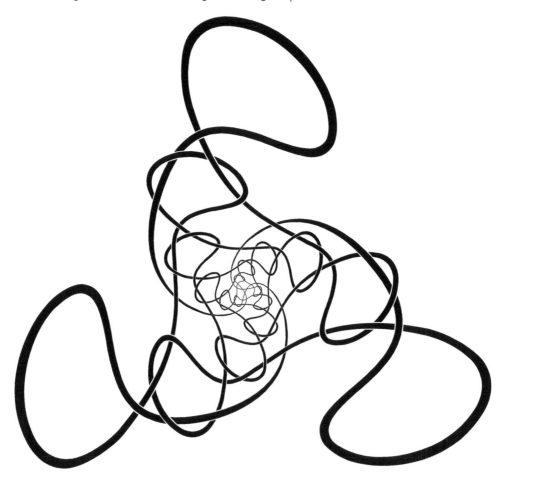

Figure 18.9. A knot based on the decorated tiles in Figure 18.8. Loose ends were joined in the middle and around the edges to create a single strand. The edges of the tiles are not shown, so only the knot is visible.

a single color. This is the same fractal tiling shown in Figure 9.16.

Applying knot strands to fractal tilings generally results in links rather than knots. In the example of Figure 18.11, strand graphics have been applied to an isosceles right triangle. This tile was used to form the fractal tiling of Figure 9.10. After constructing the *f*-tiling, additional rendering was done in Photoshop to create the final design shown in the figure.

Another approach to generating iterated knots is through substituting a patch of tiles for individual tiles in the patch. In the example of Figure 18.12, the canonical drawing of the knot 8_{18} (Figure 18.12a) is rearranged to fit a hexagonal footprint (Figure 18.12b). An analogous patch of tiles is shown in Figure

18.12c. The entire tile patch is substituted for each hexagon in Figure 18.12d. If the knot of Figure 18.12b is considered to be a decorated regular hexagon, a similar substitution yields the knot of Figure 18.12e. Iterating yields the knot of Figure 18.12f. The connections were carefully chosen to keep the knot to a single strand at each iteration. An arbitrarily complex knot can be generated by continued iteration, just as an arbitrarily large patch of tiles can be generated by continued substitution.

Another example is shown in Figure 18.13, where the canonical drawing of the trefoil knot (Figure 18.13a) is rearranged to fit a circular footprint (Figure 18.13b). While it may not look like the same knot, a series of Reidemeister moves can turn one into the

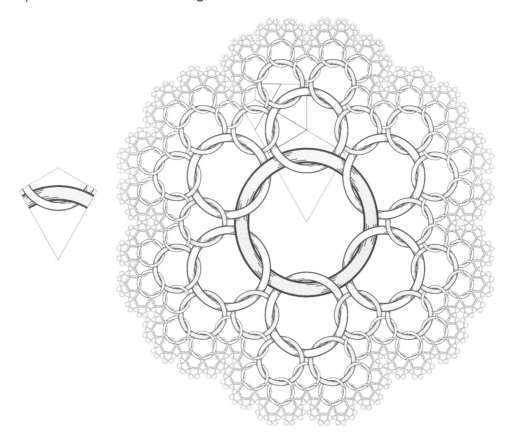

Figure 18.10. Applying ring graphics to a kite-shaped tile to create a fractal arrangement of rings.

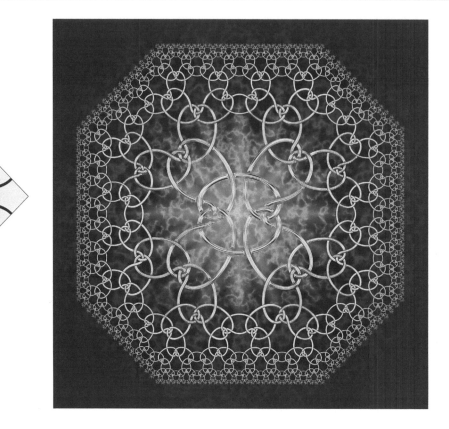

Figure 18.11. Knot graphics were applied to an isosceles right triangle tile to create this fractal design.

other, which proves they're the same knot mathematically. An analogous patch of tiles is shown in Figure 18.13c, and the entire patch is substituted for each circle in Figure 18.13d. If the knot of Figure 18.13b is considered to be a decorated circle, a similar substitution yields the knot of Figure 18.13e. An infinite number of iterations would yield a fractal knot, the basis for the digital artwork entitled "Infinity" (2006), shown in Figure 18.13f. Additional examples of iterated knots constructed in a like manner can be found in Fathauer 2007.

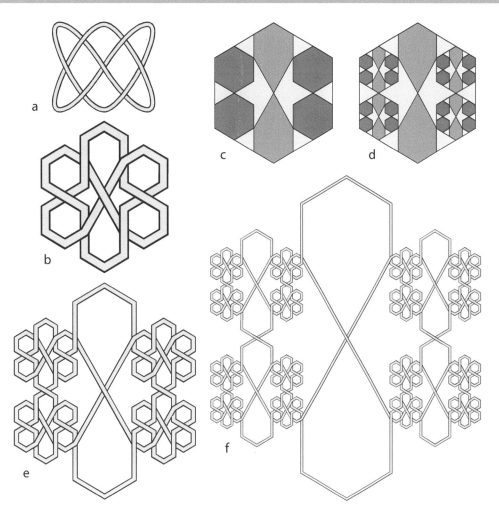

Figure 18.12. (a) A canonical drawing of the knot 8_{18}. (b) A self-similar version of the same knot. (c) An analogous patch of tiles. (d) A patch of tiles where the hexagons in the starting patch have been substituted with the whole starting patch. (e) A knot iterated through a similar substitution. (f) A second iteration of the knot.

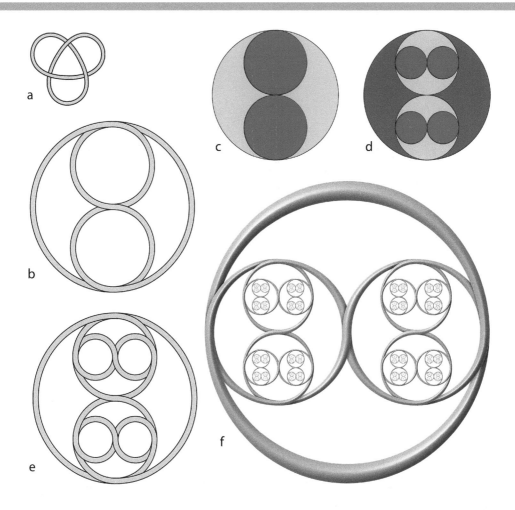

Figure 18.13. (a) A canonical drawing of the trefoil knot. (b) A self-similar version of the same knot. (c) An analogous patch of tiles. (d) A patch of tiles where the circles in the starting patch have been substituted with the whole starting patch. (e) A knot iterated through a similar substitution. (f) An artwork entitled "Infinity" that depicts the knot after an infinite number of iterations.

Other types of decorative graphics

A wide variety of decorations can be applied to tiles. As an example, spiral arms were used to decorate regular polygons in Figure 18.14. Regular polygons can tessellate in many different ways, as detailed in Chapter 2. Two different patches of tiles created using the tiles of Figure 18.14 are shown in Figure 18.15, with some color swaps. If the original tile edges are ignored each colored region can be considered to form one tile. In this case, each tile is a double-armed spiral, with each end being a triangular, square, or hexagonal spiral.

Another example, in which decoration of tiles is used to create a unique design that no longer looks like a tiling, is shown in Figure 18.16. The f-tiling used is a flower-like arrangement of six $s = 6$ pods of the sort shown in

Figure 18.14. Three regular polygons decorated with spiral graphics.

Figure 18.15. Two patches of tiles constructed from the tiles of Figure 18.14, with some colors changed to ensure each colored region is a double-armed spiral.

Figure 18.16. Graduated coloring of the triangular tiles in a fractal tiling. The edges of one of the first-generation tiles are outlined in black.

Figure 9.10. An invariant mapping was used to apply a graduated color to each tile in a way that it matches smoothly across tile boundaries [Ouyang and Fathauer 2014].

Activity 18.1. Creating symmetrical designs by decorating tessellations

Materials: Copies of Worksheet 18.1 on card stock, scissors or craft knife, pencil, pen, eraser, and crayons or colored pencils (or other means of marking and coloring card-stock tiles).

Objective: Learn to create a symmetrical design by decorating tiles and fitting them together.

Vocabulary: Decorated tile, patch of tiles.

Activity Sequence

1. Write the vocabulary terms on the board and discuss their meanings.
2. Pass out copies of the worksheet and other materials.
3. Have the students decorate the two top tiles with graphics that meet seamlessly when mated to other square and triangular tiles. They should use at least one feature that crosses tile edges. They can use lines, dots, geometric shapes like stars, real-world objects like flowers, etc. It's fine to leave some white space around the features they draw. Before they start, they should decide how they want the patterns they draw to match up. Their completed designs should have some sort of symmetry: translational, rotational, and/or reflection.
4. Walk around the room and watch how the students decorate their top two tiles, to make sure they have tiles that will tessellate seamlessly.
5. Have the students decorate the remaining tiles on their sheets the same way as the top two.
6. Have the students cut out their decorated tiles and experiment with different ways of fitting them together to form a patch of tiles with an overall symmetrical design.

Discussion Questions

1. How did you decorate your tiles? After cutting the tiles out and assembling them, did you think you made a good choice? How would you change it if you could do it over again?
2. Do you think using decorated tiles is an easier way to make a symmetrical design than just drawing a full design on a sheet of paper? Why or why not?
3. Which of the various tiles your classmates created do you like the best? Why?
4. What characteristics make for a good design of this sort?

Worksheet 18.1. Creating symmetrical designs by decorating tessellations

Decorate the two top tiles with graphics that will meet seamlessly when mated to other square and triangular tiles. You may want to use a mixture of features that cross tile edges and features that don't. Before you start, decide how you want the features you draw to match up. For example, if you want to put four squares together at the center of a design that has four-fold symmetry, then the graphics along edge A should match up to the graphics along edge B. The tick marks are included to guide the alignment of features.

When you are satisfied with your decoration of the labeled tiles, make multiple copies of each one on the unlabeled tiles. Then cut them out and assemble them in a patch to form a design. Try rearranging your tiles to form different designs.

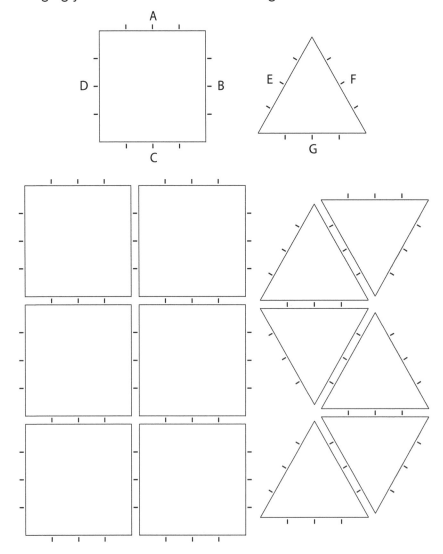

Chapter 19

Tessellation Metamorphoses and Dissections

A **metamorphosis, or morph** *for short, is a change from one thing to another. A tessellation morph is a tessellation that evolves spatially or temporally.* Escher's *Metamorphosis* prints evolve linearly through multiple transitions from one tessellation to another [Bool et al. 1982]. These artworks are engrossing and have inspired many others to try their hand at tessellation metamorphoses, whether geometric or Escheresque. One of my morph designs is shown in Figure 19.1.

In addition to spatial morphs, temporal morphs are possible. In Escher's time, animation was a laborious process, but today it's relatively easy to create simple animations. As a result, there are many entertaining geometric tessellation morphs that can be viewed online. Obviously these can't be shown in a book, but the interested reader may wish to look at the examples referenced later in the chapter.

Geometric metamorphoses

An example of a geometric morph is shown in Figure 19.2. Three common tessellations of regular and star polygons can be seen in the middle and on either end. Some intermediate columns of tiles make for a relatively smooth transition between these tessellations. In this example, the morph takes place in one direction. Morphs in more than one direction are possible but are more difficult to create.

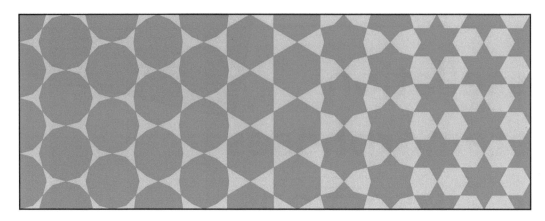

Figure 19.1. A morph from, left to right, tessellations of dodecagons and three-pointed stars to hexagons and triangles to six-pointed stars and hexagons.

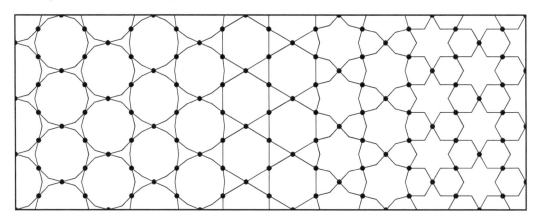

Figure 19.2. The vertices of the tessellation morph of Figure 19.1, marked here as black dots, form a lattice that doesn't change over the morph.

Different techniques that can be employed in tessellation morphs are discussed below. One simple approach is to keep the lattice formed by the vertices unchanged over the morph. This was done in Figure 19.2, as illustrated in Figure 19.3. In this case, creating the morph is simply a matter of distorting the tile edges between vertices, the basic process used to create an Escheresque tile.

Linear geometric morphs were explored by Williams S. Huff in the 1960s. He called them "parquet deformations" and had his architecture students design them [Hofstadter 1986]. Craig Kaplan has developed schemes to allow parquet deformations to be generated by computer [Kaplan 2010].

Many people have created animated geometric tessellation morphs. Dave Whyte has posted a number of fine examples on Twitter under #beesandbombs [Whyte 2020]. Software such as GeoGebra can be used to facilitate the creation of geometric animations, including tessellation morphs. Examples can be seen on the page of the Facebook group "GeoGebra Arts & STEAM" [GeoGebra 2019] and on the Twitter feed of Vincent Pantal [Pantal 2020].

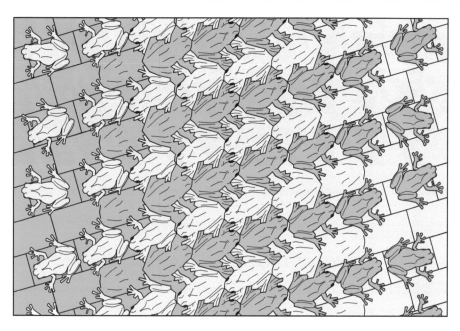

Figure 19.3. A tessellation metamorphosis which transitions from realistic frogs to a tessellation of stylized frogs to realistic frogs again.

Positive and negative space

A key tool used by Escher to make his metamorphoses effective is the manipulation of positive and negative space. **Positive space** *is a graphics design term that denotes the portion of a design perceived as being in the foreground.* **Negative space** *denotes the portion of a design perceived as being in the background. Positive forms are referred to as* **figure***, while the negative space surrounding them is referred to as* **ground***.

Escher reverses our perception of positive and negative space in many of his prints. An example from my own work is shown in Figure 19.4, originally designed as a commission for a T-shirt for the Chicago Marathon. On the left, the black runners are perceived as positive

figures against a negative white ground. In the middle is a very abstract tessellation of black and white runners. On the right, white runners are perceived as positive figures against a negative black ground. As the eye scans over the print, the brain switches its interpretation of which color is ground and which is figure. In a sense, this is a type of optical illusion and is typical of the sort of game Escher enjoyed playing with our perception of two-dimensional prints. As an artwork, I think it's more compelling than Figure 19.3, because it tells a story. The evolution of the runner's form from left to right resonates with a marathon runner's mental state from the start to the middle to the finish of the race. Similarly, many of Escher's best prints not only evoke worlds but tell stories.

Figure 19.4. "Marathon" (2004), a print in which the figure and ground colors flip from one side to the other.

Techniques for transitioning between Escheresque tessellation motifs

Escher's woodcut "Metamorphosis II", begun in 1939 and finished in 1940, is a monumental work and is nearly 4 meters long, printed from 20 blocks. Several devices were employed to effect the transitions between tessellations, and in fact, there are several sections of the print that are not tessellations. A description of the work follows, with different devices enumerated. The bulk of the print is shown in Figure 19.5, with both ends left off.

Starting at the left end, text on a neutral background (not a tessellation) becomes the edges of a tessellation of squares (device #1). Next a two-color tessellation of squares becomes a two-color tessellation of lizards. This is accomplished by distorting the edges of the tiles without moving the vertices (device #2), as described for Figures 19.1 and 19.2. In parallel, the interior decoration of the tiles (the addition of interior lines, etc.) is used to further suggest the lizard motif (device #3).

By distorting a lifelike motif, in this case lizards, Escher then transitions from a square

grid to a hexagonal grid (device #4). It's remarkable how quickly and effortlessly this transition is done. After distorting edges to turn the lizards into hexagons, Escher uses a real-world example of a tessellation of hexagons, a honeycomb, to introduce a new lifelike motif, bees (device #5). The bees grow larger and more stylized until they tessellate with fish (device #6).

A figure/ground reversal then creates fish that don't tessellate (device #7). The fish become more abstracted until they tessellate with birds, which become non-tessellating birds via another figure/ground reversal. These birds then change shape until they tessellate with two other birds on a hexagonal lattice, with each hexagon divided into three 60°/120° rhombi. One of the two bird motifs starts from an empty space and grows gradually until they match up with the first set of birds, with the third set of birds coming from the left-over space (device #8).

A perspective illusion is then used in which the mind interprets 60°/120° rhombi as cubes seen in perspective (device #9). The rhombi/cubes are then modified gradually into a small Italian town on the Amalfitani coast. A tower from the town extends into the

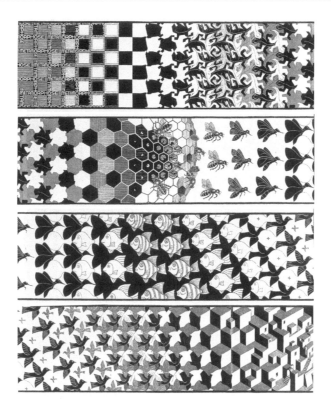

Figure 19.5. A portion of "Metamorphosis II", a woodcut by M.C. Escher.

water (not shown in Figure 19.5), where it can be interpreted as a rook on a chess board (a similar device to the hexagon tessellation to honeycomb transition). This allows a return to the checkerboard tessellation and then the words from which the print started at the other end.

An analysis of this print thus reveals that Escher employed a number of different mechanisms that he had developed over time to create this masterpiece of tessellation art. Some of these devices were used in two more tessellation morphs I made. In the top part of Figure 19.6, birds are released from a tessellation in a similar manner to Escher's 1955 lithograph "Liberation". In Figure 19.7, a morph is used to allow a transition between two rigid objects, which otherwise present problems as subjects for tessellations due to the inflexibility of their outlines. In a morph such as this, where there is no extended section of tessellating motifs, it could be argued that the word tessellation shouldn't be used. It's very much in the style with some portions of "Metamorhphosis II", however.

Figure 19.6. A tessellation morph going from musical notes to birds.

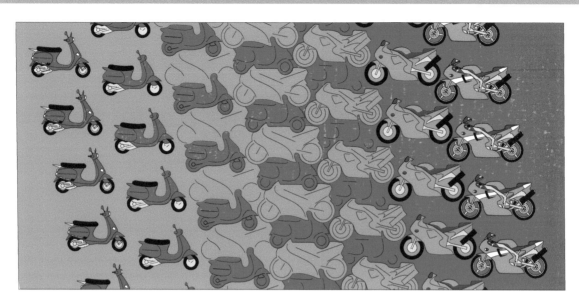

Figure 19.7. A tessellation morph from scooters to superbikes.

Tessellation dissections

A geometric dissection is the cutting of a geometric shape or shapes into pieces that can be reassembled to form a second geometric shape or shapes. One method for designing geometric dissection is superposing two tessellations that have the same translation unit cell area and the same translation vector [Frederickson 1997]. This technique actually creates a tessellation of geometric dissections rather than just a single geometric shape.

An example is shown in Figure 19.8, where a regular tessellation of squares is laid over a tessellation of squares of two different sizes. This results in a dissection "puzzle", consisting, in this case, of a set of five pieces that can form either one or two squares. The square grid could be translated relative to the underlying tessellation to create different dissections of the same general sort.

A related technique for designing geometric dissections is the superposition of two strip tessellations [Frederickson 1997]. In Figure 19.9, a single row from each of regular triangle and square tessellations are overlaid on one another. They were scaled so that the area of one square and one triangle are the same. In addition, the angle between the two was chosen such that the two sides of the square strip cross two consecutive downward-pointing triangles at the identical points A. The square strip can be translated to form different dissections, but a symmetric position was selected in Figure 19.9, which results in the same cuttings for the downward- and upward-pointing triangles. The resulting dissection is the well-known triangle-to-square dissection that allows a hinged transformation between the two shapes [Frederickson 1997].

Taking this design approach to the extreme, I overlaid four tessellations to create a set of 23 tiles that can be arranged to form an equilateral triangle, a square, a hexagon, or an octagon (Figure 19.10). This dissection was used in a puzzle called Polygony™.

The technique of superposing tessellations can be applied to Escheresque tessellations

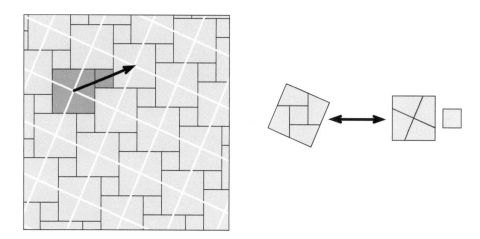

Figure 19.8. Overlaying a regular tessellation of squares on a tessellation of squares of two different sizes, under the condition that the two tessellations share a translation vector and have the same size translation unit cell, creates a dissection from one square to two squares.

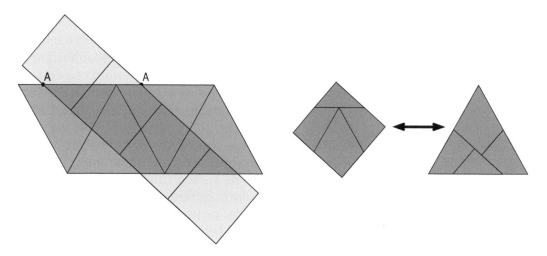

Figure 19.9. Overlaying a strip tessellation of equilateral triangles with a strip tessellation of squares to create a four-piece dissection puzzle.

as well as geometric ones. In Figure 19.11, two very different tessellations of fishermen and fish are successfully superposed as the translation unit cells have the same area and the two tessellations have a common translation vector. In a single translation unit cell of the superposed tessellations, ten tiles form a dissection that can either be arranged to form one fisherman or three fish. This design was used to produce a physical puzzle called "Fisherman's Catch", using bilayer foam that is blue on one side and orange on the other.

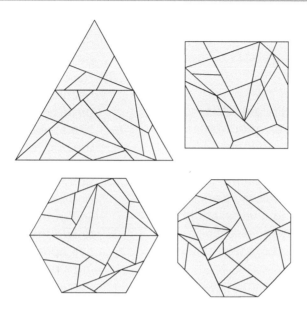

Figure 19.10. A set of 23 tiles that will form four different regular polygons.

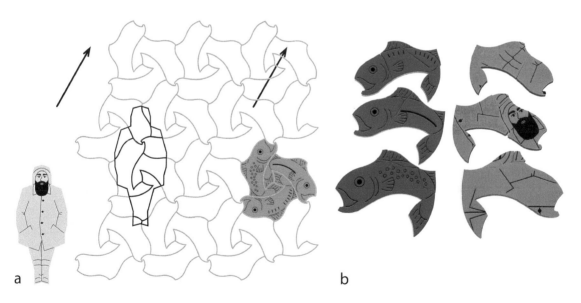

a   b

Figure 19.11. (a) Superposition of tessellations of fishermen and fish. This generates a ten-piece dissection whose tiles can either be arranged to form a fisherman or three fish. (b) Physical puzzle pieces made of bilayer foam, assembled as fish and flipped over to show both sides.

Activity 19.1. Learning to draw a tessellation metamorphosis

Materials: Copies of the Square Tile Grids (Figure 13.4) on card stock and the Square Tessellation Grid (Figure 13.5) on a paper that can be drawn and colored on, scissors or craft knife, pencil, pen, eraser, and crayons or colored pencils (or other means of coloring a tessellation).

Objective: Learn to create a tessellation morph by hand, based on a tile that fits a square grid.

Vocabulary: Metamorphosis (morph), tessellation metamorphosis.

Activity Sequence

1. Write the vocabulary terms on the board and discuss their meanings.
2. Pass out the copies and other materials.
3. Have the students create a square-based Escheresque tile using the Square Tile Grids. The easiest template for this is 13.1 or 13.3. If 13.3 is used, it should be rotated to 90°, so the left and right edges are the same and the top and bottom edges are related by a reflection.
4. Orienting the Square Tessellation Grid sheet lengthwise (landscape orientation), have the students draw a column of squares at the left edge by connecting grid dots with straight lines. Then have them draw a column of their Escheresque tiles at the right edge of the sheet. If

Template 13.1 or 13.3 was used, the right and left lines of their tiles will be identical.

5. Have the students draw lines connecting the dots in the remaining three columns to create the tile edges of a tessellation morph. First connect the two remaining columns of dots vertically with tile edges that transition from the straight line of the second column of dots to the modified line of the fifth column of dots. Then draw three lines connecting consecutive horizontal dots, also transitioning smoothly from the straight lines connecting the first and second dots to the modified lines connecting the fifth and sixth dots.
6. Have the students complete their tessellation morph by adding interior details in a way that transitions gradually from open squares to Escheresque tiles. They can then color their tessellation morph.

Discussion Questions

1. What template did you use for your tile and what motif? Do you think it was a good choice for a morph? How would you change it if you could do it over again?
2. What step did you find the most challenging? Why?
3. Which of the various morphs your classmates created do you like the best? Why?
4. What characteristics make for a good design of this sort?

Chapter 20

Edge Face

Vertex

Introduction to Polyhedra

Polyhedra are to three dimensions what polygons are to two dimensions. Recall that a polygon is a closed planer figure made up of straight-line segments. In a regular polygon, all of the line segments are of the same length, and all of the interior angles are the same. A convex polygon is one that contains all the line segments connecting any two of its points, and polygons that are not convex are concave.

Basic properties of polyhedra

A **polyhedron** is a geometric solid, the faces of which are polygons. A **face of a polyhedron** is thus a flat region of polygonal shape. An **edge of a polyhedron** is a straight-line segment shared by two adjacent faces. A **vertex of a polyhedron** is a point where three or more faces meet (Figure 20.1). A **convex polyhedron** is one for which any line connecting two points on the surface lies in the interior or on the surface of the polyhedron. All the polyhedra in this book are convex except for the stella octangula. Euler's polyhedron formula gives a simple relationship between the number of vertices V, edges E, and faces F for a convex polyhedron: $V - E + F = 2$.

The angle between two faces of a polyhedron is known as the **dihedral angle**. The **dual of a polyhedron** is the polyhedron obtained by connecting the midpoints of each face with their nearest neighbors (Figure 20.2). The dual of the dual is the original polyhedron. For example, the octahedron and cube are duals of one another.

b

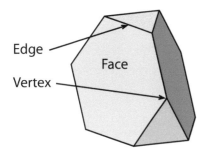

Figure 20.1. Examples of a face, an edge, and a vertex of a polyhedron.

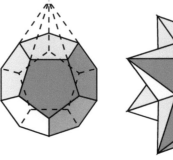

Figure 20.4. Extending the faces of a dodecahedron to form a stellated dodecahedron. Note that the colors of the faces are consistent for the regular and stellated versions.

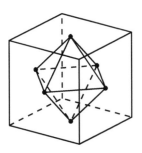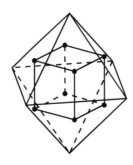

Figure 20.2. The cube and octahedron are duals of one another.

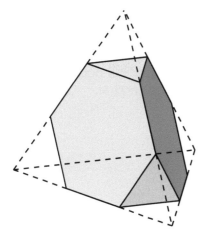

Figure 20.3. Truncation of a regular tetrahedron (dashed lines) to form a truncated tetrahedron.

Undecorated net models of many polyhedra are provided in Chapters 22–24. A variety of facts about the polyhedra are printed on these – the name, the number and type of faces, the number of vertices, the number of edges, the polyhedron's dual, and its dihedral angle(s).

*A **truncated polyhedron** is one that can be formed by cutting off portions of another polyhedron.* For example, a truncated tetrahedron can be formed by cutting off the points of a tetrahedron, as shown in Figure 20.3. The canonical way of choosing how much to cut off is to choose an amount that makes all edges the same length. *A **stellated polyhedron** is created by extending the faces of a polyhedron to form a new polyhedron that is star-like.* Figure 20.4 shows a dodecahedron with the extensions above one face along with the full stellated dodecahedron.

The five **Platonic solids** are the only convex polyhedra for which each face is the same regular polygon and all vertices are of the

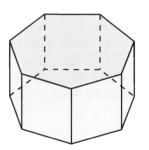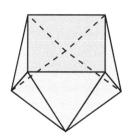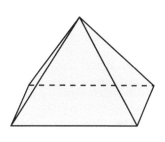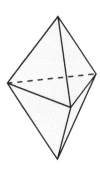

Figure 20.5. From left to right, a heptagonal prism, a square antiprism, a square pyramid, and a triangular dipyramid.

same type. They were known before Plato, but he and his students studied them. They are discussed further in Chapter 22.

There are 13 **Archimedean solids**, *convex polyhedra for which each face is a regular polygon, with more than one type of regular polygon, and for which all vertices are of the same type.* Drawings of all 13 appear in Johannes Kepler's 1619 book *Harmonices Mundi.* They are described in more detail in Chapter 23.

The names of polyhedra can be intimidating. These are made up of smaller building blocks, some of which are: tetra, 4; hexa, 6; hepta, 7; octa, 8; dodeca, 12; icosi, 20; triaconta, 30; rhombi, related to a rhombus; and hedron, faces. An example of their usage is "rhombic triacontahedron", which is a polyhedron with 30 rhombus-shaped faces.

There are many other types of polyhedra. Some of the simpler ones are prisms, antiprisms, pyramids, and dipyramids. *A (regular right)* **prism** *has two congruent regular polygon faces separated by a band of squares.* A cube is a square prism. *In an* **antiprism**, *the regular polygons are rotated with respect to one another so that a band of triangles (equilateral if the antiprism is regular) can connect them.* An octahedron is a triangular antiprism. *A (regular right)* **pyramid** *consists of a regular polygon (the "base") and triangles meeting at a common vertex (the "apex"), located at a point that lies along a line including the centroid of the base and perpendicular to the base (forming a right angle).* A dipyramid is formed by two pyramids joined base-to-base. In addition to being a triangular antiprism, the octahedron is a square dipyramid. Examples of these polyhedra are shown in Figure 20.5.

Polyhedra in art and architecture

Polyhedra were studied by mathematicians and artists alike in the Renaissance. For artists, learning to draw polyhedra helped them master perspective. For some thinkers, polyhedra symbolized religious or philosophical truths. An example of their use is a stellated dodecahedron surrounded by hexagonal prisms included by Paolo Uccello in a mosaic (ca. 1430), on the floor of the Basilica of San Marco in Venice (Figure 20.6). Leonardo da Vinci prepared numerous illustrations of polyhedra, and many of these appear in Luca Pacioli's book "De Divina Proportione" (ca. 1500). Two examples, the dodecahedron and the cuboctahedron, are shown in Figure 20.7. As best we know, Leonardo was the first person to depict polyhedra as frames rather than solids, allowing the viewer to see the structure of the backside as well as the front.

Albrecht Dürer's 1514 masterpiece "Melencolia I" prominently features a polyhedron (Figure 20.8). Johannes Kepler included drawings of polyhedra, such as that shown in Figure 20.9, in his 1619 book *Harmonices Mundi*. Early in his career, he associated the Platonic solids with the orbits of the planets.

Figure 20.6. A stellated dodecahedron used by Uccello as an architectural decoration in the 15th century.

Figure 20.7. Drawings of polyhedra by Leonardo da Vinci. (Scans from a reproduction of Pacioli's book were cleaned up a bit for this figure).

Figure 20.8. "Melencolia", by Albrecht Dürer, features a polyhedron.

M.C. Escher incorporated polyhedra in several of his prints, including "Waterfall" (1961), "Gravity" (1952), and "Double Planetoid" (1949). Salvador Dalí's 1955 painting "The Sacrament of the Last Supper" includes a portion of a dodecahedron arching over the table, and his 1954 painting "Crucifixion (Corpus Hypercubus)" depicts the crucified Jesus upon the three-dimensional net of a four-dimensional hypercube.

Polyhedra have been a rich source of inspiration for contemporary geometric sculpture. Sculptors whose work relies heavily on polyhedra and related forms include George Hart [Hart 2019], Ulrich Mikloweit [Mikloweit 2019], and Rinus Roelofs [Roelofs 2019]. One of my own ceramic sculptures, the geometric basis for which is a Goldberg polyhedron with 146 faces, is shown in Figure 20.10.

For practical reasons, most buildings are based on cubes and cuboids, with right angles. Pyramids and prisms have been used from early on to make buildings stand out. The square pyramids of the ancient Egyptians

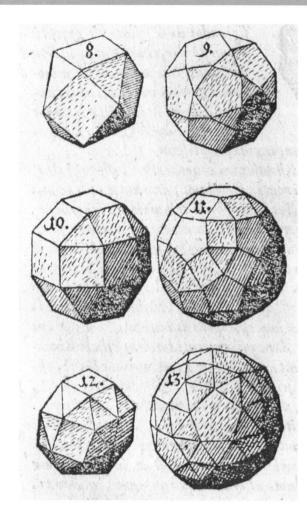

Figure 20.9. Polyhedra drawn by Johannes Kepler, from *Harmonices Mundi*.

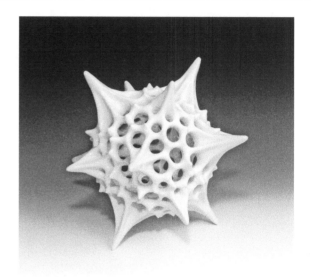

Figure 20.10. "Radiolarian Form", a ceramic sculpture by the author (2018).

are among the most famous structures in the world. There are numerous buildings that are shaped like prisms, and octagonal prisms seem particularly popular. An ancient example is the Tower of the Winds in Athens, constructed in the 1st or 2nd century BCE. In modern architecture, a wider variety of polyhedral forms is seen. For example, geodesic domes, made famous by Buckminster Fuller, are composed of triangular faces.

Polyhedra in nature

In nature, polyhedra are frequently observed in rocks and minerals. This is a result of the underlying geometric arrangement of atoms in crystal lattices. Minerals that exhibit cubic forms are the most common and include sodium chloride, or table salt. Some minerals exhibit octahedral as well as cubic forms, including pyrite and fluorite. Pyrite can also exhibit irregular dodecahedra. A fluorite specimen containing a dense jumble of cubic forms is shown in Figure 20.11.

There are many examples of polyhedra on the microscopic scale, including individual molecules. Carbon 60, a fullerene or buckyball, has the form of a truncated icosahedron with a carbon atom at each vertex. The outer protein shells of viruses can also have polyhedral shapes such as the icosahedron. In addition, there are radiolarians with icosahedral skeletons and pollens with polyhedral forms (Figure 20.12).

It could be argued that the form of the pollen grain in Figure 20.12 is a spherical tessellation rather than a polyhedron. Due to the fact that organic objects are not geometrically perfect, there's no clear dividing line between the two. A **spherical polyhedron** *is a projection of a polyhedron onto the surface of a sphere.* Spherical polyhedra, such as those shown in Figure 10.7, are a type of spherical tessellation. Not all spherical tessellations are spherical polyhedra, though. For example, the spherical tessellations of Figure 10.8 cannot be turned into polyhedra.

Figure 20.11. Fluorite specimen exhibiting cubic forms.

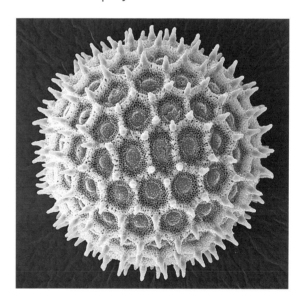

Figure 20.12. An electron micrograph of a pollen grain, taken by Louisa Howard at the Dartmouth College Electron Microscope Facility.

Tiling three-dimensional space

Just as certain polygons can tessellate the plane, certain polyhedra can tessellate three-dimensional space. *A tessellation in three or* more dimensions is usually referred to as a **honeycomb**, *with 3-honeycomb referring to a tessellation of three-dimensional space.*

We're all familiar with how cubes can be stacked without gaps to fill an arbitrarily large volume. None of the other Platonic solids will tessellate space by themselves (when not allowing fractal arrangements), but tetrahedra and octahedra together will (Figure 20.13). The fact that we live in three dimensions allows us to look down on a plane and see everything at once. We don't have this ability when viewing three-dimensional objects, so it's harder to see what such a tessellation would look like. In Figure 20.14, two other highly symmetrical polyhedra that will tessellate space are shown. One is an Archimedean solid, the truncated octahedron (Figure 20.14a). The other is the rhombic dodecahedron (Figure 20.14b).

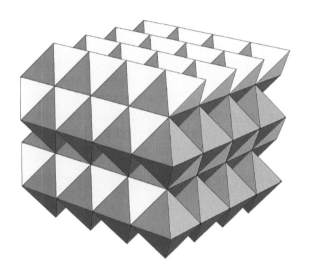

Figure 20.13. Octahedra and tetrahedral together will tessellate space.

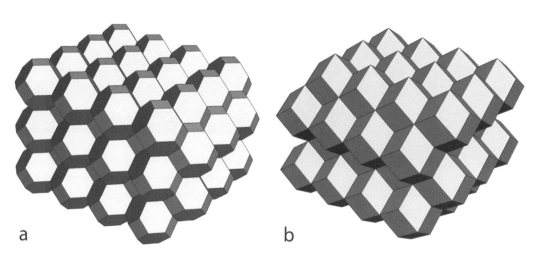

a

b

Figure 20.14. (a) Truncated octahedra and (b) rhombic dodecahedra will tessellate space.

Slicing 3-honeycombs to reveal plane tessellations

It's entertaining and enlightening to consider how slicing 3-honeycombs can reveal two-dimensional tessellations. Consider an edge-to-edge tiling of space by cubes, with alternate cubes given a second color, a sort of 3D analog of a checkerboard. A 4 × 4 × 4 block of cubes of this sort (a patch of 3D tiles) is shown in Figure 20.15a. Each face of the block reveals a regular tessellation of squares. Slices at different angles reveal different tessellations, and it may come as a surprise that a regular tessellation of triangles and a semi-regular tessellation of triangles and hexagons can be revealed in this way (Figure 20.15b–d).

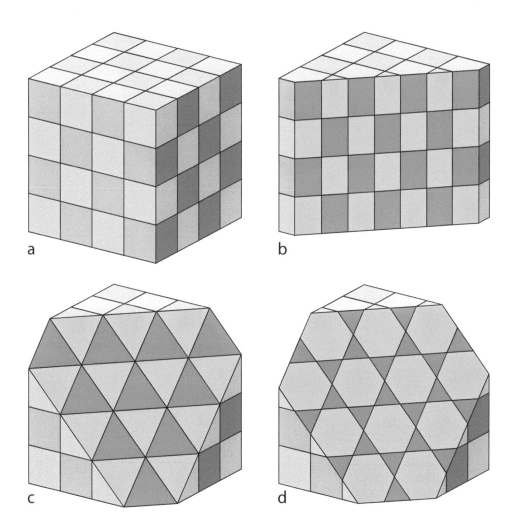

Figure 20.15. A space-filling two-color arrangement of cubes: (a) unsliced; (b) sliced to reveal a tessellation of rectangles; (c) sliced to reveal a tessellation of equilateral triangles; (d) sliced to reveal a semi-regular tessellation of triangles and hexagons.

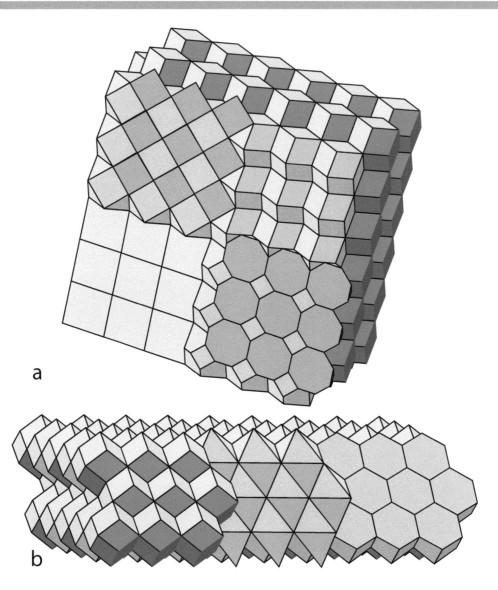

Figure 20.16. A space-filling arrangement of rhombic dodecahedra can be sliced to reveal (a) regular square and semi-regular square-and-octagon tessellations and (b) regular triangle and hexagon tessellations.

As another example, a 3-honeycomb of rhombic dodecahedra can be sliced to reveal all three regular tessellations as well as the square-and-octagon semi-regular tessellation, depending on orientation and depth of the slice (Figure 20.16).

Activity 20.1. Identifying and characterizing polyhedra in nature

Materials: Copies of Worksheet 20.1.

Objective: Learn to recognize polyhedra in nature and also to characterize their properties.

Vocabulary: Polyhedron, dodecahedron, octahedron, vertex, edge, face.

Specific Common Core State Standard for Mathematics addressed by the activity

Use geometric shapes, their measures, and their properties to describe objects (e.g., modeling a tree trunk or a human torso as a cylinder). (G-MG)

Activity Sequence

1. Pass out the worksheet.
2. Write the vocabulary terms on the board and discuss the meaning of each one.
3. Describe how pyrite can exhibit cubic, octahedral, or dodecahedral crystals. Note that the dodecahedra aren't regular due to the crystallography, though they may appear so. Have the students draw at least three dodecahedra on the photograph and then have them share their results.
4. Have the students draw a dot at each vertex on the pinecone photo, considering the individual scales as faces of a polyhedron. Then have them connect the dots with lines that will represent the edges of a polyhedron. Have them share their results for the number of faces meeting at each vertex and the number of edges possessed by each face.
5. Have the students draw dots where three or more flowers meet. In some locations, one dot could be drawn between four flowers or two dots close together with three flowers meeting at each dot. Neither way is incorrect. Have the students connect the nearest neighbor dots to draw a polyhedron that approximates the shape of the flower cluster. Have them calculate the average number of faces meeting at each dot and the average number of edges for each face. Then have them choose which of the three polyhedra shown most nearly approximates their drawing. Have them share their results and contrast two cases where students drew different collections of dots.

Worksheet 20.1. Identifying and characterizing polyhedra in nature

Polyhedra play an important role in many objects found in nature. In some cases, it's relatively easy to recognize them. In others, more thought is required to recognize the polyhedron and understand its properties.

Photograph (a) shows a pyrite ("fool's gold") specimen that exhibits dodecahedral crystals. Draw at least three full dodecahedra on the photo that incorporate some faces of the crystals. Below, the photo (b) are three different views of regular dodecahedra to help guide your drawings.

The pinecone shown in (c) can also be modeled as a polyhedron. Draw a dot at each vertex; i.e., point where three or more of the individual scales meet. Don't worry about including points near the edges of the photo for which only portions of the plates are visible. Now connect the dots to their nearest neighbors. The lines you draw will be the edges defining faces that approximately correspond to the boundaries of the individual plates. How many faces meet at each vertex? How many edges do the faces have?

The photo (d) shows a small cluster of flowers. Draw a dot at each vertex; i.e., point where three or more of the individual flowers meet. The locations of the points is not as clear as for the pinecone, but do your best. Now connect the dots to their nearest neighbors. The lines you draw will define faces of a polyhedron that models the overall shape of the flower cluster. What is the average number of faces meeting at each vertex? What is the average number of edges possessed by each face? Of the three polyhedra (e), below the photo (d), which most nearly approximates the polyhedron you drew? Why?

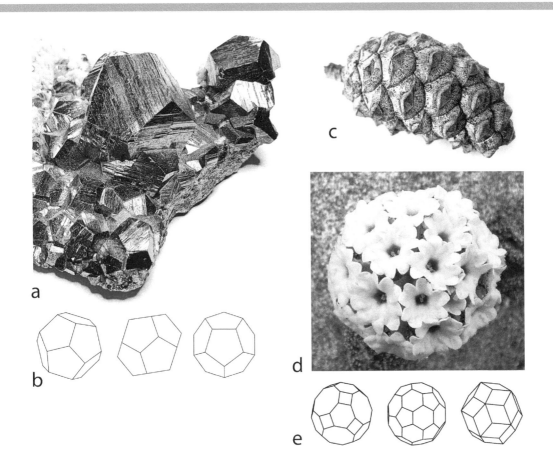

a

b

c

d

e

Activity 20.2.
Identifying polyhedra in art and architecture

Materials: Copies of Worksheet 20.2 for students.

Objective: Learn to recognize polyhedra in art and architecture, and gain a better appreciation and understanding of why and how they are used.

Vocabulary: Platonic solid, Archimedean solid, pyramid, prism.

Specific Common Core State Standard for Mathematics addressed by the activity

Use geometric shapes, their measures, and their properties to describe objects (e.g., modeling a tree trunk or a human torso as a cylinder). (G-MG)

Activity Sequence

1. Pass out the worksheet.
2. Write the vocabulary terms on the board and discuss the meaning of each one.
3. Have the students complete the first part of the worksheet, on the painting of Luca Pacioli. Have them share their results.

   The large solid in the painting is a small rhombicuboctahedron, with three squares and one equilateral triangle meeting at each vertex. The small solid is a dodecahedron, with three pentagons meeting at each vertex. Some possible reasons for including polyhedra in the painting are: to remind the viewer of who Pacioli was and what he studied; as an homage to Pacioli's accomplishments as a mathematician, which included publication of an influential text on

geometry that included numerous drawings of polyhedra; and as a way for the artist to demonstrate his mastery of perspective drawing.

4. Have the students complete the second part of the worksheet, on one of the Giza Pyramids.

   The structure is a square pyramid, or even more specifically, a right square pyramid (since the apex is directly above the centroid of the base), with one square and two triangles meeting at four vertices and four triangles meeting at the other vertex. Some possible reasons for using a square pyramid for this structure are the fact that a polyhedron is a bold and attention-grabbing form and is of religious significance of the pyramid in Ancient Egypt.

5. Have the students complete the third part of the worksheet, on the Tower of the Winds.

   This building is in the form of an octagonal prism, or more specifically a right octagonal prism (since the angles formed by the sides and base are right angles), with one octagon and two rectangles meeting at each of the 16 vertices. Some possible reasons for using an octagonal prism for this structure are: a prism lends itself to a tower form, since the side rectangles can be made as tall and narrow as desired; the eight sides pay tribute to the eight wind deities; and the correspondence of the sides with the compass points would make it easier for a person looking up at the wind vane to determine which direction the wind was blowing.

Worksheet 20.2. Identifying polyhedra in art and architecture

Polyhedra are found in paintings, sculpture, and buildings from ancient times to the present.

Figure (b) shows a late 15th-century painting of Luca Pacioli, along with close-ups of two polyhedra found in the painting. Describe the types of vertices found in each of the two polyhedra. Write the names of the two polyhedra. The larger one is one of the Archimedean solids shown in (a), while the smaller one is a Platonic solid. Pacioli was an Italian mathematician, Franciscan friar, and collaborator of Leonardo da Vinci. Why do you think the polyhedra might be included in the painting?

Figure (c) shows famous polyhedral structures that were built as monuments to Egyptian rulers. Describe the number and types of vertices found in one of these structures. Write the name of the polyhedron, being as specific as possible. What possible reasons can you think of to explain why this polyhedral form was chosen for these structures?

Figure (d) shows a famous polyhedral structure from the Roman Empire. This structure, known as the "Tower of the Winds", was topped with a weather vane that indicated the direction of the wind. The building has eight sides, and Romans had eight wind deities that corresponded to the compass points north, northeast, east, ... Describe the number and types of vertices found in the polyhedron that corresponds to the overall shape of the building. Write the name of the polyhedron. What possible reasons can you think of to explain why this polyhedral form was chosen for this structure?

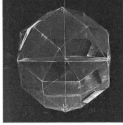

a   Cuboctahedron   Small rhombicuboctahedron   Great rhombicuboctahedron   Snub cube

c

Photograph by Robster1983 [Wikipedia 2019].

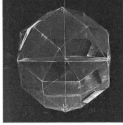

b

d

Photograph by Joanbanjo [Wikipedia 2019].
Cropped and contrast adjusted.

Chapter 21

Adapting Plane Tessellations to Polyhedra

Plane tessellations can be adapted to cover the surface of polyhedra. This chapter describes additional restrictions imposed by the geometry of the polyhedron, as well as different tricks and techniques that allow plane tessellations to be used for the decoration of polyhedra. Subsequent chapters go through details specific to particular polyhedra. One appeal of this activity is the fact that a tessellation on the surface of a polyhedron is finite in extent but has no boundaries. Handling a physical model is like holding an infinite tessellation in the palm of your hand.

67.5

60° 60°

60°

hexagon

Nets of polyhedra

A **net of a polyhedron** is a drawing show-ing the constituent polygons laid out flat and joined at their edges in a fashion that would allow it to be folded up to form the polyhedron. Different nets exist for a given polyhedron, depending on how the polygons are joined. Figure 21.1 shows some of the 11 distinct nets that will form a cube.

In Chapters 22–25, the nets are given for a number of polyhedra, both decorated with tessellations (designated with an "a" after the number) and without decoration (designated "b"). Some of the tessellated versions are line drawings that can be colored, and some are shown colored. The undecorated versions include information about the polyhedra. If you would like a completely blank polyhe-dron, you can assemble it inside out (with the print on the inside). The nets in these chapters include tabs to facilitate the assembly of the polyhedra.

Restrictions on plane tessellations for use on polyhedra

The concept of translation in a plane doesn't apply in a straightforward fashion to the surface of a polyhedron, so translational sym-metry and glide reflection symmetry don't play an explicit role in designing tessellations for the surfaces of the polyhedra. Rotational symmetry and mirror symmetry are important for polyhedra. Since a polyhedron is a three-dimensional object, these concepts need to be modified to make sense in three dimen-sions. Rather than considering rotation to be about a point, it is considered to be about a line. Reflection is considered to occur about a plane rather than a line.

In general, applying plane tessellations to polyhedra is relatively challenging. Most tes-sellations cannot be applied to a given polyhe-dron. The problem lies in the angular measure

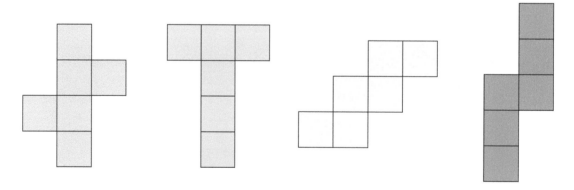

Figure 21.1. Four distinct nets that will form a cube.

of the faces meeting at a vertex. In a plane, this measure is always 360°. At the vertex of a convex polyhedron, however, it is less than 360°. For a cube, for example, the angles meeting at a vertex sum up to 270°. In order for a plane tessellation to be amenable to use on a cube, it would need to have four points at the four corners of a square that had the property that the tessellation could be wrapped around the corners of a cube at each of those points and still match up. A point in a tessellation about which the tessellation was identical every 90° would have that property; in other words, a point of four-fold rotational symmetry.

Considering the tessellation of Figure 21.2, there are a few different squares that can be drawn with a point of four-fold rotation at each corner (actually infinitely many if the squares are just made larger and larger). Each of these squares will be one face of a cube. The purple square will result in three seahorses meeting at each vertex of the cube, the red square results in three eels meeting at each vertex, and the

black results in a cube with three seahorses at four vertices and three eels at the other four. Strictly speaking, a net of a cube needs to be drawn on the tessellation in a way that each vertex of the net is a point of four-fold rotational symmetry. Since each square in Figure 21.2 is part of a lattice of similar squares, that requirement is fulfilled for any net of a cube.

Unfortunately, most polyhedra are even more difficult than a cube, because their faces are not all the same type of polygon.

Distorting plane tessellations to fit polyhedra

One approach to creating a tessellation for the surface of a polyhedron is to design a new tessellation specifically for that polyhedron. Templates are provided in later chapters to make that easier, as described in the next section. In many cases, though, it's possible to distort an existing tessellation to fit a polyhedron. Two examples of this technique are described in this section.

In the first example, the starting point is a tessellation with kaleidoscopic symmetry. This is one for which every edge of a polygonal tile is a line of mirror symmetry. In Figure 21.3, each tile is an equilateral triangle, and these appear in six orientations. Note that every vertex is a point of three-fold rotational symmetry. The tile could easily be applied to a polyhedron in which every face is an equilateral triangle and every vertex has even valence (a regular octahedron), but it would fail at any vertex with odd valence.

By distorting the tile, it can also be used to decorate a more complex polyhedron such as a snub cube. In a snub cube, there are squares as well as equilateral triangles. Four equilateral triangles and one square meet at

Figure 21.2. Three different ways to position a square on a tessellation of seahorses and eels to allow its application to the surface of a cube.

Figure 21.3. A tessellation with kaleidoscopic symmetry.

each vertex, so the angles sum up to 330°. This plane tessellation can be fit to the snub cube by distorting two of the six triangles to accommodate the 30° difference between 360° and 330°. This is accomplished by distorting an equilateral triangle tile to become an isosceles right triangle tile. Two of the corners of the triangle decrease in angle by 15° to accommodate the 30° lost at the vertices

of the polyhedron, and four of these distorted triangles are fit together at their 90° corners to form a square (Figure 21.4). Applying the tessellation to a snub cube involved making three distinct squares, with different motifs along the edges in each. Distorting the triangle is a simple operation in a drawing program such as Illustrator. However, the distortion of the motifs doesn't look very good in this case, and

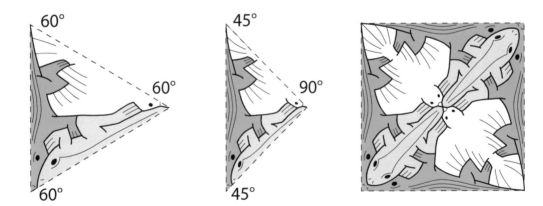

Figure 21.4. Distortion of a decorated equilateral triangle tile to fit an isosceles right triangle, allowing construction of a decorated square tile.

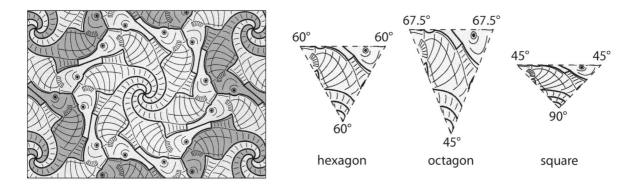

Figure 21.5. Distortion of the equilateral triangle tile of a seahorse tessellation to allow the formation of decorated hexagons, octagons, and squares.

additional reshaping of the boundaries of the motifs and interior details results in a better looking tessellation, as seen in Chapter 23.

In the second example, a tessellation of seahorses will be distorted to fit a great rhombicuboctahedron. This polyhedron has an octagon, a hexagon, and a square meeting at each vertex. The tessellation of Figure 21.5 has points of six-fold symmetry where the tails meet and points of three-fold symmetry where the heads meet. Two modified versions are created, with eight seahorses meeting at the tails for the octagons and four seahorses meeting at the tails for the squares.

Three heads, one from each of these three versions, will meet at each vertex of the great rhombicuboctahedron.

In the plane tessellation, a full seahorse occupies one equilateral triangle (one sixth of a hexagon), though it's chopped up in pieces. For the octagonal faces, a full seahorse needs to occupy one eighth of the face, which is a triangle with two angles of 67.5° and one of 45°. For the square, two angles of 45° and one of 90° are required. Making these distortions by simply stretching the equilateral triangle, as done for Figure 21.5, results in unsatisfactory motifs in the octagon and square, as seen on the left in

Figure 21.6. Octagon and square tiles formed by distorting an equilateral triangle from the tessellation of Figure 21.5, before and after reshaping features to improve the seahorse motifs.

Figure 21.6. Significant reshaping of the lines was done to improve the motifs (right in Figure 21.6) for use in the great rhombicuboctahedron in Chapter 23. Another approach to improving the distorted motifs would be to use a more sophisticated mapping that ensured the angles of features, such as the seahorse tail, matched more smoothly at tile boundaries.

Designing and drawing tessellations for polyhedra using the templates

Templates are provided at the end of Chapters 22–25 that can be used to design tessellations for the various polyhedra. Some expertise in designing plane tessellations is recommended before attempting to design tessellations for polyhedra, as the templates for the polyhedra are relatively complex compared to simpler templates based on single squares and triangles (Chapters 13 and 15). If you haven't designed plane tessellations before, you may wish to work with some of the plane-tessellation templates earlier in this book to help you develop this skill.

The templates in the following chapters are sized to fit the nets. In some cases, the templates provided are the ones used to create the tessellation for that polyhedron in the book, but in many cases they aren't. To use a template, first make copies of the page to work on. An example of a template is shown in Figure 21.7. Each letter denotes an independent line segment. Each instance of the same letter denotes an instance of the same line segment. The orientation of the L-shaped line next to the letter indicates the orientation of the line segment for that instance, and the large arrows show the transformation from one instance to another. For example, there are two instances of line segment (a) in the template in Figure 21.7, and the two are related by a 180° rotation about the midpoint of the left edge of the isosceles triangle. Together these two make up the full left edge.

Once you've designed a tessellation motif using the template, you will need to tile the net for the polyhedron using multiple copies of your motif. In cases where the use of the template might not be clear, an example is given, as shown in Figure 21.7 for the octahedron. In addition to the outline of your tile, you will want to draw any interior details in each instance of the tile. You might want to cut out and score the tiled net, then fold and check the assembly before laying it flat again to color it before the final assembly.

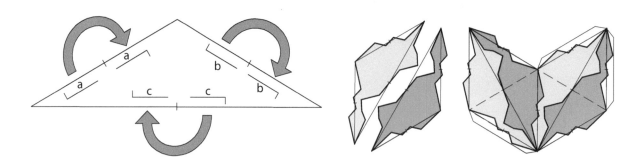

Figure 21.7. An example of the sort of template presented in the following chapters to guide design of Escheresque tessellations for polyhedra.

Coloring of tessellations on polyhedra

There are several considerations and choices for coloring the faces of a polyhedron [Hart 2019]. For polyhedra decorated with an Escheresque tessellation, there may be more than one tile per face, adding another dimension to the problem. The coloring of decorated polyhedra is a matter of personal taste, but some aspects of the problem will be discussed here. The polyhedra can be colored using anything that marks on paper – colored pencils, felt-tip markers, crayons, paints, etc. In general, you will get better results by coloring the nets on a flat surface rather than coloring the assembled polyhedra due to the fact that better control of the pencil, etc., will be possible. The downside to coloring the flat net is that it can be challenging to visualize which tiles will be adjacent once it's folded up. For this reason, it is a good idea to do test foldings of the polyhedron at various stages as you color.

The simplest way to color the polyhedra is to color all instances of a given motif the same. Examples of this sort of coloring are found in the seahorses and eels on the cube (polyhedron number 2) and the alligators, tadpoles, and birds on the snub cube (polyhedron number 15).

Another popular way to color tessellations is to make adjacent tiles different colors. Adjacent tiles are tiles sharing an edge. M.C. Escher was fond of tessellations that only required two colors, which he usually chose to be black and white in his finished prints. Any tessellation, planar or otherwise, for which every vertex has even valence is two-colorable. Considering the undecorated Platonic solids, only the octahedron is two-colorable. Three colors are required for the cube and icosahedron and four for the tetrahedron and dodecahedron. Any tessellation or polyhedron with a single vertex that has odd valence is not two-colorable. The first two Archimedean solids in Figure 21.8 allow two coloring of their faces, while the other two don't. The flower design on the small rhombicuboctahedron is two colored (polyhedron number 10, as is the flower design on the small rhombicosidodecahedron, which is polyhedron number 12). An example of a tessellated polyhedron requiring three colors is the octopus tessellation on the truncated cube (polyhedron number 14).

Rather than simply requiring that adjacent tiles be of different colors and then using the minimum number of colors required to achieve this, it can be more satisfying from a design standpoint to color the individual tiles in a way that is sympathetic to the symmetry

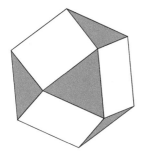 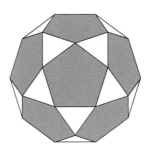  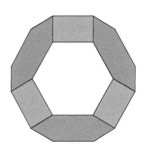

Figure 21.8. Two two-colorable Archimedean solids and two requiring three colors if adjacent faces are to have different colors.

of the polyhedron. This can be tricky for some of the more complicated Archimedean solids and can require several colors. As an example, consider the icosahedron without decoration. Groups of four of the 20 faces can be colored with five different colors such that the coloring is chiral and each group of four lies in the planes of a tetrahedron [Hart 2019].

An example of coloring a decorated polyhedron that required some thought is the lizard tessellation on the rhombic triacontahedron (polyhedron number 23). Five difference colors are used, with every group of five lizards meeting at a vertex containing one lizard of each color. There are 12 lizards altogether of each color, and for each color, they are distributed uniformly around the polyhedron. The polyhedron can be oriented such that six of the rhombic faces lie in the faces of an enclosing cube. If this is done, then each of those six rhombi contains four halves of lizards in the same four colors.

Tips on building the models

This section contains some tips to help you achieve the most pleasing assembled models without unnecessary effort. The first step is to make copies of the page(s) with the net for the polyhedron you wish to build. For a paper size of 8-1/2″ × 11″, regular copy or printer paper can be used, but it will result in a relatively flimsy model that is easily dented or crushed. A thicker paper such as card stock will be more robust. On the other hand, a thinner paper cuts and folds more easily. If you scale up a model and print on a larger paper stock, such as 11″ × 17″, you should definitely use a heavy paper such as card or cover stock. If you are printing a colored polyhedron, you will get the best color and finish using a photographic paper, but some of these are more difficult to

crease sharply. Photographing pages with a smart phone and printing the photo should work reasonably well as long as you are careful to photograph straight on in order to minimize distortion of the image.

For the nets decorated with my tessellation designs, some are in black and white and some are in color. If you don't have access to a color copier or printer and don't mind cutting up your book, you can cut the color nets out of the book. Any polyhedron that requires multiple copies of the page is in black and white.

If you are drawing your own tessellation or other design on the polyhedron, it's probably best to do so before cutting, as the paper will lay flat and allow your pencil or pen marks to extend past the edges of the faces. It would be easier to color the polyhedron before cutting as well, but in some cases you will want to fold it up before coloring to see how the tiles fit together. You can also color a little, fold enough to check your coloring, lay it flat again, color some more, fold to check it, etc.

In order to achieve crisp folds, it is best to score the fold lines (dashed). This is particularly important for heavier papers. For scoring, you should use a straight edge and some implement that will create a deep groove in the paper without breaking through. A ball point pen that no longer has ink is ideal for this purpose. Just draw it along the dashed line as if you were tracing the line, but press firmly. Scoring the backside of the sheet will result in a somewhat crisper fold, but this has the disadvantage that it's difficult or impossible to see the dashed lines. The paper can be held against a window to show the lines better, but it's awkward to work that way. A light box would be a better solution, if you happen to have one available.

Cutting can be done with scissors or with an X-Acto or utility knife. If you use a knife, it may be helpful to use a straight edge. Also be sure and use a thick backing such as mat board to avoid cutting your work surface. The more accurately you score and cut your polyhedron, the better the finished product will look.

Assembling the polyhedra is not too difficult for the simple ones but can be relatively challenging for some of the models with a large number of faces. The easiest way to get them together is probably to use tape, but this will not look very good. Using adhesive on the tabs such as glue stick or white paper glue will give a better result. Holding the model in place while the tabs dry can be tedious. You may find it helpful to use a tape that can be cleanly removed later on, such as blue painters tape (similar to masking tape, but better quality).

If you wish to hang your finished model, you may want to tie a paperclip on a string or fishing line and seal the paperclip inside the polyhedron as you assemble it, with the bulk of the string remaining outside.

Activity 21.1. Representing solids using nets

Materials: Copies of Worksheet 21.1 for the students.

Objective: Learn to use nets to represent polyhedra and solve problems.

Vocabulary: Polyhedron, net, prism.

Specific Common Core State Standard for Mathematics addressed by the activity

Represent three-dimensional figures using nets made up of rectangles and triangles, and use the nets to find the surface area of these figures. Apply these techniques in the context of solving real-world and mathematical problems. (6.G 4)

Activity Sequence

1. Pass out the worksheet.
2. Write the vocabulary terms on the board and discuss the meaning of each one.
3. Have the students cross through the nets that they think wouldn't work for the room and draw a net of their own. Ask a couple of students to share their nets.
4. Have the students calculate the area of the room. Ask a couple of students to share their answers.
5. Have the students calculate how many gallons of paint would be needed to paint the room. Ask a couple of students to share their answers.

Worksheet 21.1. Representing solids using nets

The net of a polyhedron is a drawing of all the faces laid out flat in a way that would allow them to be folded up to form the polyhedron, as shown in the figure for the cube.

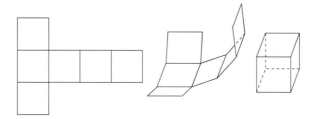

Imagine you are given the job of painting a storage room that has two walls at right angles and a third wall forming an isosceles right triangle with the other two. The room is therefore in the shape of a triangular prism. The net for a given polyhedron can be drawn in more than one way. Which of the nets shown would work for the room? Cross through the one(s) that wouldn't work.

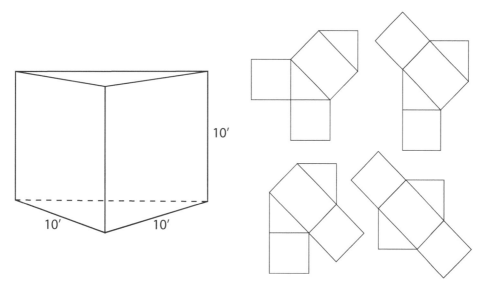

Draw at least one more net that would work. Using your net to help you, along with the Pythagorean theorem, calculate the area of the interior of the room.

If you wanted to paint the room, and you knew one gallon of paint would cover 200 square feet, how many gallons of paint would you need to buy to paint the whole room, including the floor and the ceiling?

Activity 21.2. Using transformations to apply a tessellation motif to a net

Materials: Copies of Worksheet 21.2 for the students.

Objective: Learn to apply a tessellation motif to a polyhedron net using plane transformations.

Vocabulary: Polyhedron, net, rotation, translation, transformation.

Specific Common Core State Standard for Mathematics addressed by the activity

Given a geometric figure and a rotation, a reflection, or a translation, draw the transformed figure using, e.g., graph paper, tracing paper, or geometry software. Specify a sequence of transformations that will carry a given figure onto another. (G-CO 5)

Activity Sequence

1. Pass out the worksheet.
2. Write the vocabulary terms on the board and discuss the meaning of each one.
3. Have the students draw the rest of the bird graphics in the isolated square labeled B. Ask a couple of students to share their drawings.
4. Have the students draw bird graphics in all of the squares in the net. Ask a couple of students to share their drawing.
5. Have the students answer the questions about transformations.

   There are different transformations that will create the same end result, so the following answers are not the only correct ones. To transform square 1 into square 2, rotate 90° clockwise about the lower left corner of square 1. To transform square 1 into square 3, translate to the right by one square's width and down by one square's height. To transform square 1 into square 4, rotate 90° clockwise about the lower left corner of square 1, and then translate down by twice a square's height.
6. If time allows, the students can color their nets, cut them out, fold them up, and tape or glue them to create a tessellated cube.

Worksheet 21.2. Using transformations to apply a tessellation motif to a net

A tessellated polyhedron is created by filling in each polygon in the polyhedron's net with tessellation graphics that will match up properly when the polyhedron is assembled. The net shown for a cube is to be filled with the graphics for a bird tessellation. The graphics are shown in the square labeled A. The three squares labeled A on the net will be filled just like the example. However, the B squares need to be filled with the same graphics rotated to 90°.

In the isolated square labeled B, finish drawing the bird graphics, but rotated to 90° with respect to square A. Some tick marks are included to help guide your drawing. First draw the heavy lines that outline the birds. Then draw the interior details. Once you're finished and are sure it's done correctly, make multiple copies of the graphics on the net, oriented as indicated.

The different squares in the net are related to one another by rotations and translations. Describe a transformation that will take square 1 to square 2. Be as specific and quantitative as possible.

Describe transformations that will take square 1 to square 3, and then square 1 to square 4.

When you're finished drawing the birds, you can color them. Then cut out the net along the solid lines, crease the paper along the dotted lines, and fold it up to form a cube, tucking the tabs underneath the square faces. Use glue or tape to hold the cube together.

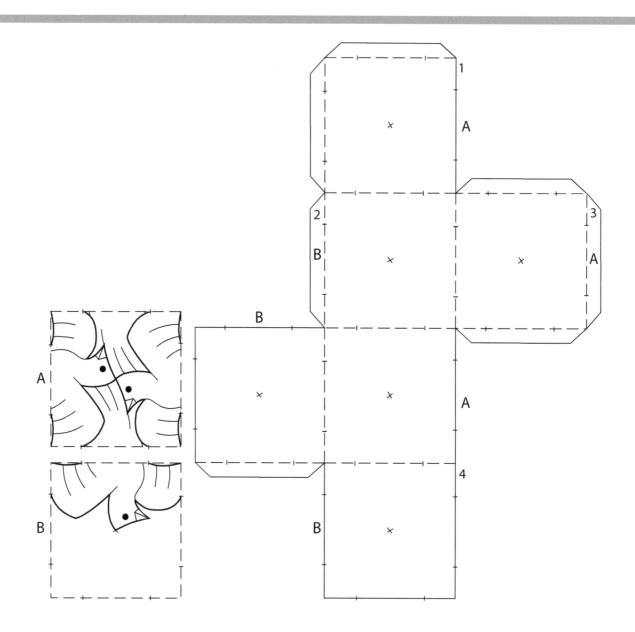

Chapter 22

Tessellating the Platonic Solids

Tetrahedron

Cube

Octahedron

Dodecahedron

Background on the Platonic solids

The five Platonic solids are the only convex polyhedra for which each face is a congruent regular polygon and all vertices are of the same type. They are named after the number of faces they possess, though the regular hexahedron is more commonly referred to as the cube (Figure 22.1).

While the Platonic solids weren't discovered by Plato, he and his students studied them and they've come to bear his name. The Ancient Greeks associated four of the solids with the four classical elements: the cube with earth, the octahedron with air, the icosahedron with water, and the tetrahedron with fire.

The remaining solid, the dodecahedron, was associated with the heavens.

Three of the Platonic solids have faces that are equilateral triangles. When three equilateral triangles meet at each vertex, the tetrahedron results. The octahedron is formed when four triangles meet at each vertex, and the icosahedron when five triangles meet at each vertex. When six equilateral triangles meet at a vertex, the group lies in a plane, so instead of a polyhedron a regular plane tessellation is formed.

In the case of squares, the cube is formed when three squares meet at each vertex. When four squares meet at each vertex, a regular tessellation of squares is formed. When three regular pentagons meet at a vertex, the dodecahedron results. Four regular pentagons

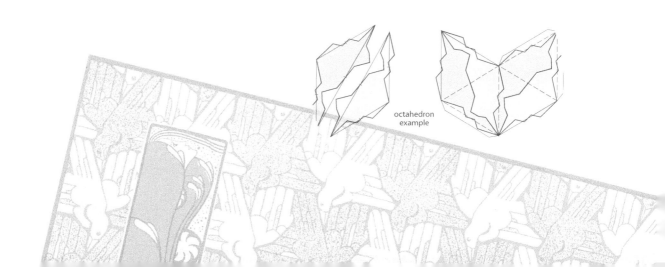
octahedron example

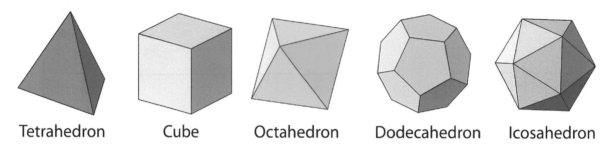

| | Tetrahedron | Cube | Octahedron | Dodecahedron | Icosahedron |

Figure 22.1. The Platonic solids.

Table 22.1. Properties of the Platonic Solids

Solid name	No. of faces	No. of vertices	No. of edges	Dihedral angle	Dual solid
Tetrahedron	4	4	6	$\approx 70.53°$	tetrahedron
Cube	6	8	12	$90°$	octahedron
Octahedron	8	6	12	$\approx 109.47°$	cube
Dodecahedron	12	20	30	$\approx 116.57°$	icosahedron
Icosahedron	20	12	30	$\approx 138.18°$	dodecahedron

would have an angular sum greater than 360°, so there is no regular tessellation based on pentagons. Finally, the meeting of three regular hexagons forms the third regular tessellation. Since any regular polygon with more than six sides has angles greater than those of the hexagon, there are no other possibilities for regular tessellations or Platonic solids. Some basic properties of the Platonic solids are summarized in Table 22.1.

On the following pages, two nets are provided for each Platonic solid, one with an Escheresque tessellation and one without. Some of the Escheresque tessellations are colored, so you can easily create a colorful tessellated polyhedron, and some are not so you can color them yourself. On the blank nets you can draw your own Escheresque tessellation with the help of the templates that follow the nets.

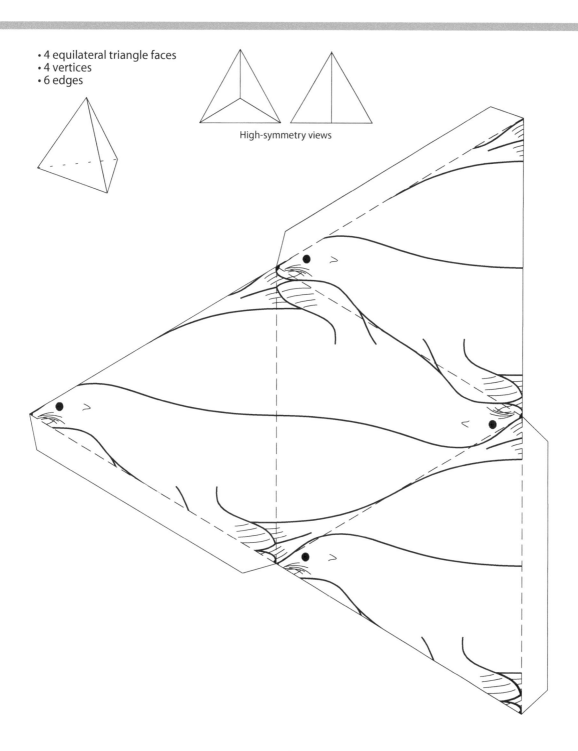

- 4 equilateral triangle faces
- 4 vertices
- 6 edges

High-symmetry views

1a. Tetrahedron with seal tessellation

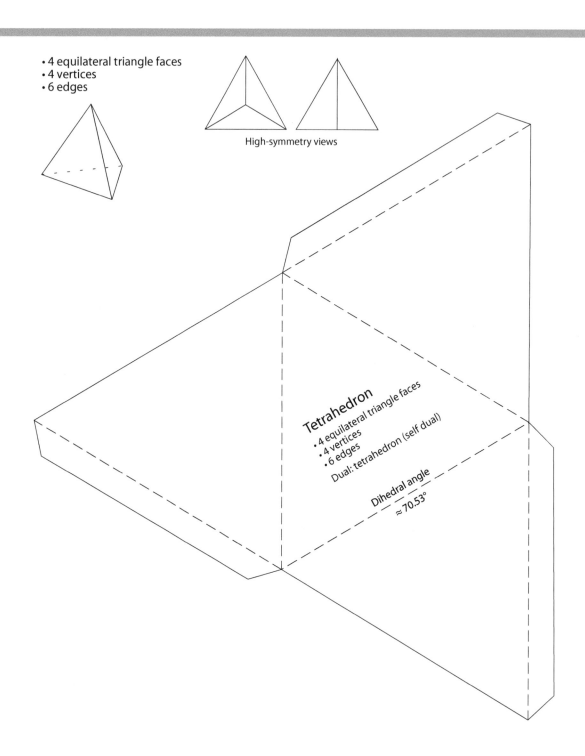

- 4 equilateral triangle faces
- 4 vertices
- 6 edges

High-symmetry views

Tetrahedron
- 4 equilateral triangle faces
- 4 vertices
- 6 edges

Dual: tetrahedron (self dual)

Dihedral angle
≈ 70.53°

1b. Tetrahedron

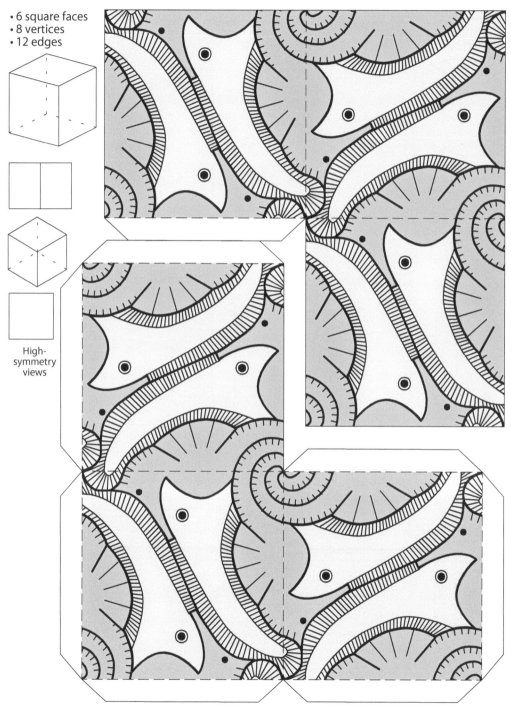

- 6 square faces
- 8 vertices
- 12 edges

High-symmetry views

2a. Cube with seahorse and eel tessellation

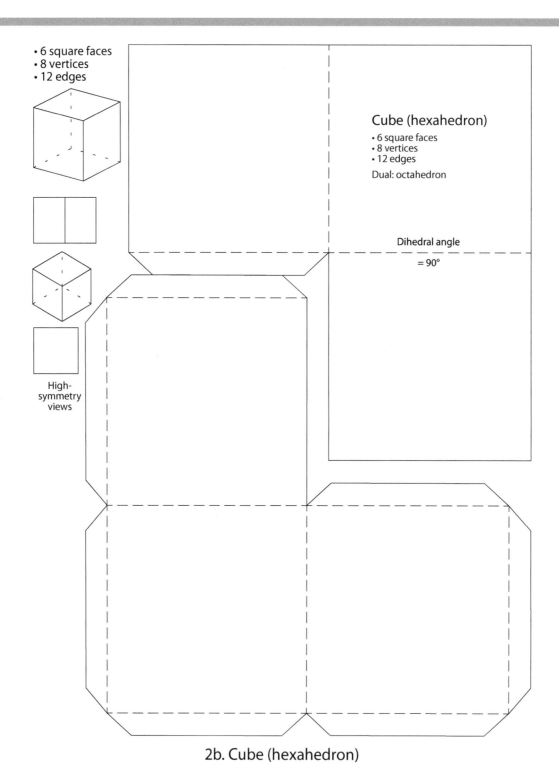

- 6 square faces
- 8 vertices
- 12 edges

High-symmetry views

Cube (hexahedron)

- 6 square faces
- 8 vertices
- 12 edges

Dual: octahedron

Dihedral angle

$= 90°$

2b. Cube (hexahedron)

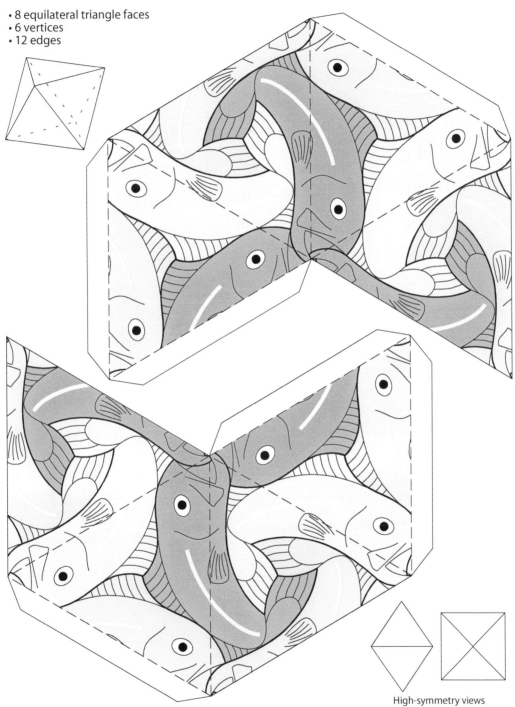

- 8 equilateral triangle faces
- 6 vertices
- 12 edges

High-symmetry views

3a. Octahedron with fish tessellation

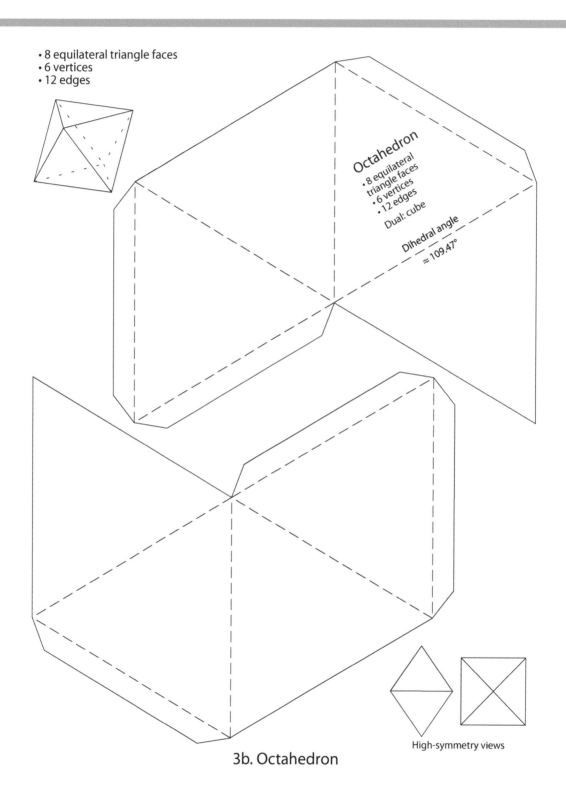

- 8 equilateral triangle faces
- 6 vertices
- 12 edges

Octahedron
- 8 equilateral triangle faces
- 6 vertices
- 12 edges

Dual: cube

Dihedral angle
≈ 109.47°

3b. Octahedron

High-symmetry views

- 12 pentagon faces
- 20 vertices
- 30 edges

Other high-symmetry views

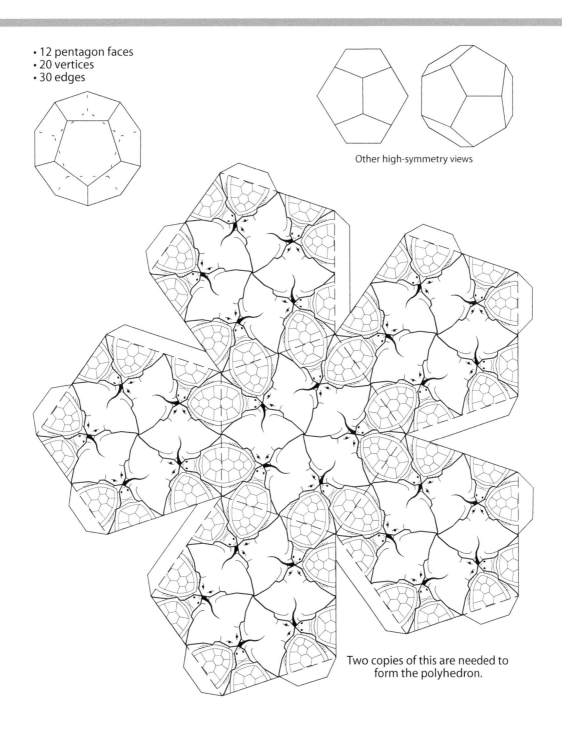

Two copies of this are needed to
form the polyhedron.

4a. Dodecahedron with ray and sea turtle tessellation

- 12 pentagon faces
- 20 vertices
- 30 edges

Other high-symmetry views

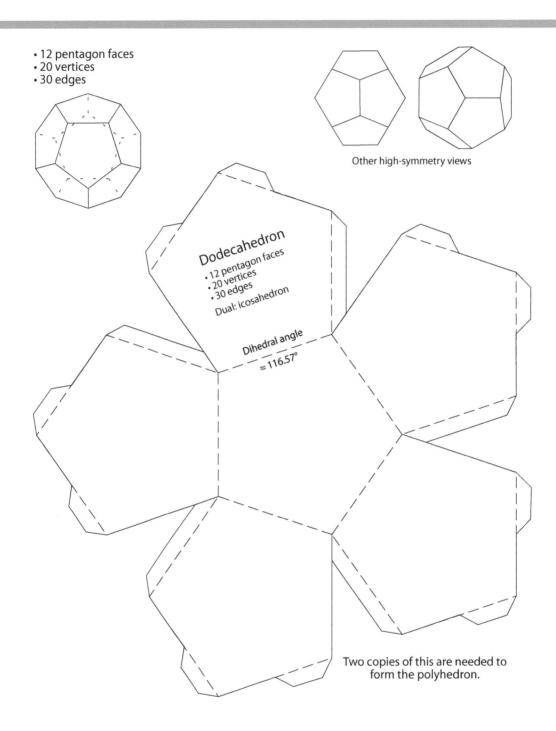

Dodecahedron
- 12 pentagon faces
- 20 vertices
- 30 edges
Dual: icosahedron

Dihedral angle
≈ 116.57°

Two copies of this are needed to
form the polyhedron.

4b. Dodecahedron

- 20 equilateral triangle faces
- 12 vertices
- 30 edges

High-symmetry views

5a. Icosahedron with goose tessellation

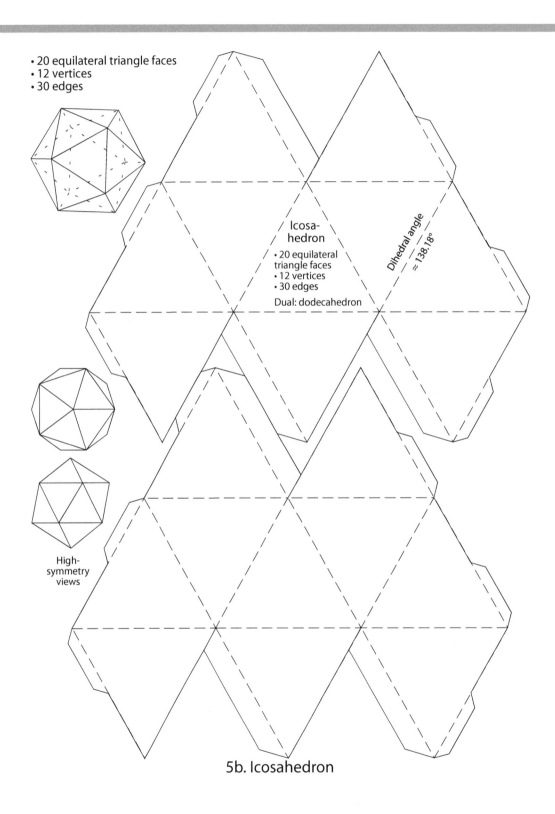

- 20 equilateral triangle faces
- 12 vertices
- 30 edges

Icosa-
hedron

- 20 equilateral
triangle faces
- 12 vertices
- 30 edges

Dual: dodecahedron

Dihedral angle
≈ 138.18°

High-
symmetry
views

5b. Icosahedron

Tessellation templates for the Platonic solids

The three templates in Figure 22.2 are isosceles triangles with angles of 30°, 30°, and 120°. There are three independent line segments a–c, each of which appears in two orientations related by a 180° rotation. This template was used for both the seals tetrahedron and geese icosahedron. It can also be used for the octahedron.

Two of these tiles, related by a 180° rotation and placed with the long edges, matched up to form a rhombus that filled two of the equilateral triangles in these polyhedra. The line segments a and b can only be different for the tetrahedron. Two narrow angles of the rhombi must be adjacent in the octahedron and icosahedron, requiring that a and b be the same.

Copy Figure 22.2 and use the appropriate template to create your own tessellation. Then tile the net for the polyhedron using multiple copies of your design, as shown in Figure 22.3 for half the octahedron. For the tetrahedron, the rhombi are oriented with the long edges of the triangles horizontal, while for the icosahedron the long edges of the triangles should be vertical.

The template in Figure 22.4 is an isosceles right triangle. There are two independent line segments a and b, each of which appears in two orientations related by a rotation. Two of these tiles fill one square face of a cube. This template was used for the sea horse and eel cube, but the template was divided into two smaller tiles. Care should be taken when applying multiple copies of this template to the cube net, as the orientation of the tiles alternates between adjacent squares.

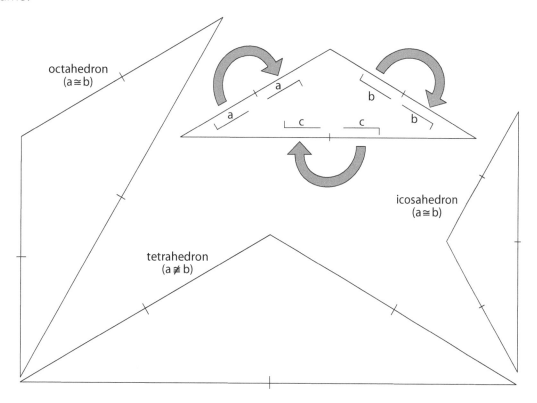

Figure 22.2. Templates to aid in designing Escheresque tessellations for three of the Platonic solids.

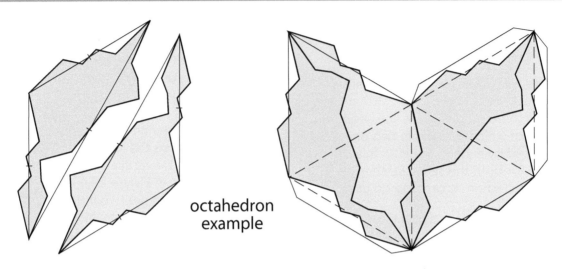

octahedron
example

Figure 22.3. How to use the octahedron template with the net to tessellate an octahedron.

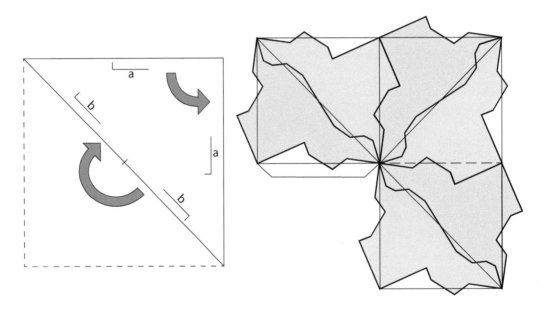

Figure 22.4. Template for designing an Escheresque tile and applying it to the net for a cube.

The template in Figure 22.5 is an isosceles triangle that constitutes 1/5 of a regular pentagon, with angles of 72°, 54°, and 54°. An example is shown applying five of these to tile one of the pentagons in the net for the dodecahedron. This is the template used for the squid and ray dodecahedron, with each tile divided into two smaller tiles.

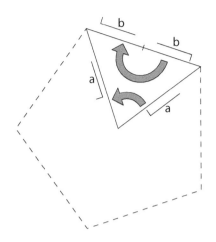
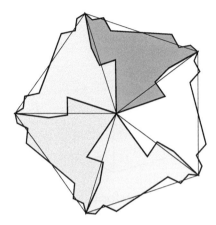

Figure 22.5. Template for designing an Escheresque tile and applying it to the net for a dodecahedron.

Activity 22.1. Attributes of the Platonic solids and Euler's formula

Materials: Copies of Worksheet 22.1 for students.

Objective: Become familiar with basic attributes of the Platonic solids, such as the number of faces, edges, and vertices possessed by each, and become familiar with and learn to apply Euler's formula, in different arrangements, relating these quantities.

Vocabulary: Platonic solid, face, vertex, edge.

Specific Common Core State Standard for Mathematics addressed by the activity
Rearrange formulas to highlight a quantity of interest, using the same reasoning as in solving equations. For example, rearrange Ohm's law V = IR to highlight resistance R. (A-CED 4)

Activity Sequence
1. Pass out the worksheet.
2. Write the vocabulary terms on the board and discuss the meaning of each one.
3. Have the students complete the table at the top of the worksheet. And then have the students share their answers.

Tetrahedron	4	6	4	3
Cube	6	12	8	3
Octahedron	8	12	6	4
Dodecahedron	12	30	20	3
Icosahedron	20	30	12	5

4. Have the students complete the next three tasks on ordering and grouping the solids. And then have the students share their results.
5. Discuss Euler's formula. Ask the students to calculate the value of C for each of the five Platonic solids, and then have them solve the remaining problems on the worksheet. Have the students share their answers.
   14, 6, 36, 60

Worksheet 22.1. Attributes of the Platonic solids and Euler's formula

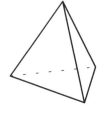 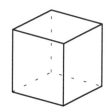 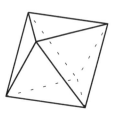 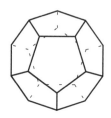 

Fill out the table below for the properties of each solid.

Name of the solid	Number of faces F	Number of edges E	Number of vertices V	Number of faces meeting at each vertex

Order the Platonic solids by the number of faces, from least to most.

Group the Platonic solids by the number of edges possessed by each polygonal face, from least to most. How many groups are there?

Group the Platonic solids by the number of faces meeting at each vertex, from least to most. How many groups are there?

Euler's formula provides a simple relationship for the number of vertices, faces, and edges in a convex polyhedron: $V + F - E = C$. Using the table above, determine the value of the constant C. Then use the formula to answer the questions below.

A cuboctahedron has 12 vertices and 24 edges. How many faces does it have?

A truncated octahedron has 24 vertices, 36 edges, and 8 hexagon faces. The remaining faces are square; how many square faces are there?

A truncated cube has the same number of faces as a truncated octahedron and 24 vertices. How many edges does it have?

A snub dodecahedron has 92 faces and 150 edges. How many vertices does it have?

Activity 22.2. Drawing the Platonic solids

Materials: Copies of Worksheet 22.2 for students.

Objective: Gain a better understanding of the Platonic solids by visualizing and drawing them from different perspectives.

Vocabulary: Perspective drawing, high-symmetry direction.

Specific Common Core State Standard for Mathematics addressed by the activity

Visualize relationships between two-dimensional and three-dimensional objects. (G-GMD)

Activity Sequence

1. Pass out the worksheet.
2. Write the vocabulary terms on the board and discuss the meaning of each one.
3. Discuss what a cube would look like from views along arrows A, B, and C. Then have the students draw the three views plus the perspective view.

Ask for volunteers to reproduce their drawings on the board or via an overhead projector.

The high-symmetry views are shown on the pages containing the net for the cube.

4. Discuss what an octahedron would look like from views along arrows A and B. Then have the students draw the two views plus the perspective view. Ask for volunteers to reproduce their drawings on the board or via an overhead projector.

The high-symmetry views are shown on the pages containing the net for the octahedron.

5. Discuss what the hidden edges would look like for the drawings of the dodecahedron and icosahedron. Then have the students draw the hidden edges as dashed lines. Ask for volunteers to reproduce their drawings on the board or via an overhead projector.

The hidden edges are shown on the pages for the dodecahedron and icosahedron.

Worksheet 22.2. Drawing the Platonic solids

Four different drawings of a tetrahedron are shown on the top line of the figure. The first is a perspective drawing, with the hidden edge indicated by a dashed line. The next two are views along the high-symmetry directions, indicated by arrows on the first drawing. The fourth is similar to the first, but omits the dashed lines and includes shading to provide more realism. Using the drawings of the cube and octahedron below as a guide, make the additional drawings indicated. On the drawings of the dodecahedron and icosahedron, add dashed lines to indicate the hidden edges.

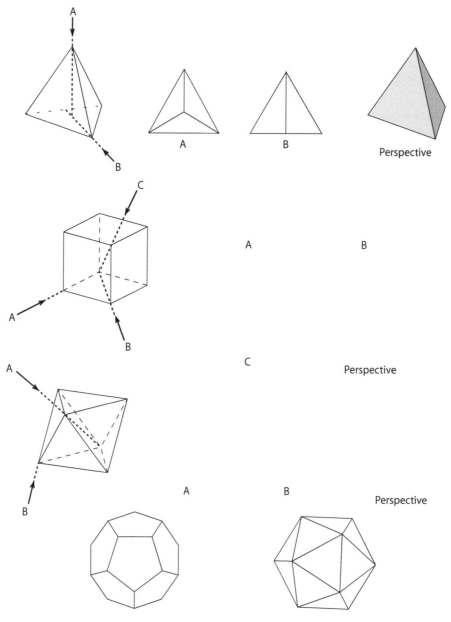

Chapter 23

Truncated tetrahedron · Cuboctahedron · Truncated octahedron · Small rhombicuboctahedron · Great rhombicuboctahedron · Small rhombicosi-dodecahedron · Great rhombicosi-dodecahedron · Truncated cube · Snub cube · Truncated dodecahedron · Snub dodecahedron · Truncated icosahedron

Tessellating the Archimedean Solids

Background on the Archimedean solids

Excluding prisms and antiprisms, there are 13 convex polyhedra for which each face is a regular polygon, but (in contrast to the Platonic solids) not all are of the same type, and for which all vertices are of the same type. These 13 solids (Figure 23.1) were known to Archimedes, after whom they are named, but his writings on them are lost. They were rediscovered in the Renaissance, with Johannes Kepler completing the list. Drawings of all 13 appear in his 1619 book *Harmonices Mundi*.

As with semi-regular plane tessellations, the Archimedean solids can be conveniently described in terms of their regular polygons meeting at each vertex. For example, the cuboctahedron, in which a triangle meets a square meets a second triangle meets a second square, has vertices described as 3.4.3.4. By convention, the smallest numbers are at the beginning (left), and the specification proceeds in a clockwise fashion.

The duals of the Archimedean solids are known as the **Catalan** solids. Since all the vertices are identical in an Archimedean solid, all the faces are identical in a Catalan solid (with possible mirrored variants). The duals and

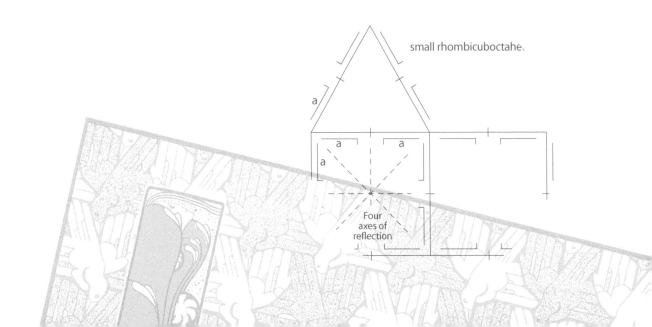

small rhombicuboctahe.

a

a · a

a

Four axes of reflection

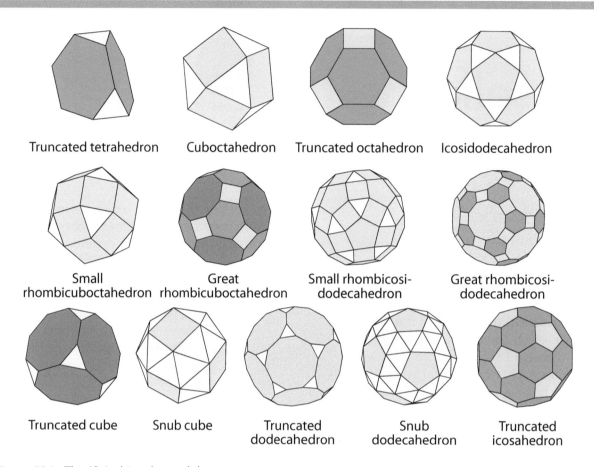

| Truncated tetrahedron | Cuboctahedron | Truncated octahedron | Icosidodecahedron |

| Small rhombicuboctahedron | Great rhombicuboctahedron | Small rhombicosi-dodecahedron | Great rhombicosi-dodecahedron |

| Truncated cube | Snub cube | Truncated dodecahedron | Snub dodecahedron | Truncated icosahedron |

Figure 23.1. The 13 Archimedean solids.

dihedral angles for the Archimedean solids are listed in Table 23.1 and are also given on the untessellated versions of the polyhedra in the following pages.

Note that the snub cube and snub dodecahedron are chiral, which means they come in distinct left-handed and right-handed versions. The left-handed and right-handed versions are mirror images of one another. The other Archimedean solids are identical to their mirror images. The truncated octahedron is the only Archimedean solid that is a space-filling polyhedron, meaning it can tessellate in three dimensions without gaps or overlaps.

Solid name	No. of vertices	No. of faces	No. of edges	Vertex description	Dihedral angles (approximate)
Truncated tetrahedron	12	8	18	3.6.6	109.47°/70.53°
Cuboctahedron	12	14	24	3.4.3.4	125.27°
Truncated octahedron	24	14	36	4.6.6	125.27°/109.47°
Icosidodecahedron	30	32	60	3.5.3.5	142.62°
Small rhombicuboctahedron	24	26	48	3.4.4.4	144.73°/135.00°
Great rhombicuboctahedron	48	26	72	4.6.8	144.73°/135.00°/125.27°
Small rhombicosidodecahedron	60	62	120	3.4.5.4	159.10°/148.28°
Great rhombicosidodecahedron	120	62	180	4.6.10	159.10°/148.28°/142.62°
Truncated cube	24	14	36	3.8.8	125.27°/90.00°
Snub cube	24	38	60	3.3.3.3.4	153.23/142.98°
Truncated dodecahedron	60	32	90	3.10.10	142.62°/116.57
Snub dodecahedron	60	92	150	3.3.3.3.5	164.18°/152.93°
Truncated icosahedron	60	32	90	5.6.6	142.62°/138.18°

Table. 23.1. Properties of the Archimedean Solids

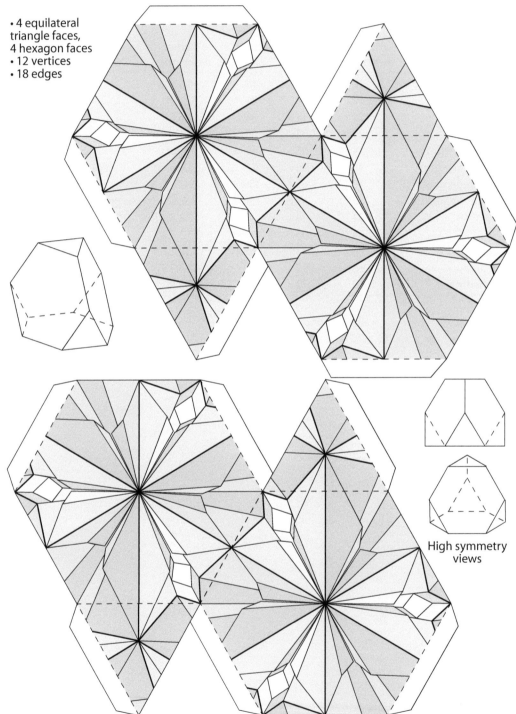

- 4 equilateral triangle faces, 4 hexagon faces
- 12 vertices
- 18 edges

High symmetry views

6a. Truncated tetrahedron with paper airplane tessellation

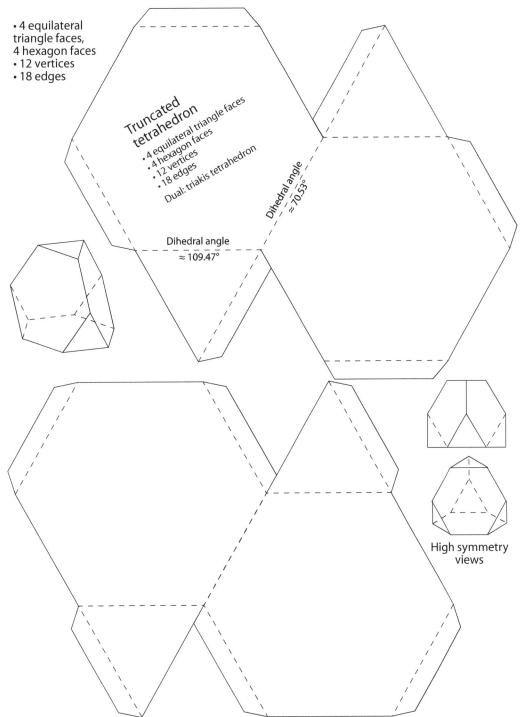

• 4 equilateral
triangle faces,
4 hexagon faces
• 12 vertices
• 18 edges

Truncated
tetrahedron
• 4 equilateral triangle faces
• 4 hexagon faces
• 12 vertices
• 18 edges
Dual: triakis tetrahedron

Dihedral angle
≈ 70.53°

Dihedral angle
≈ 109.47°

High symmetry
views

6b. Truncated tetrahedron with paper airplane tessellation

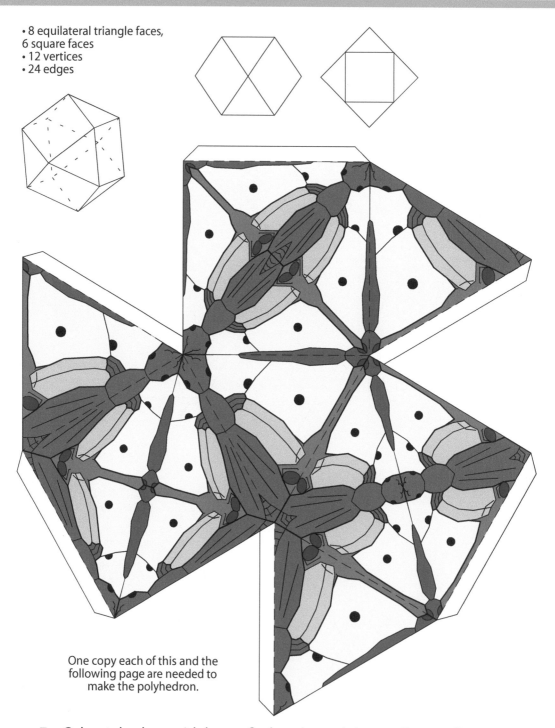

- 8 equilateral triangle faces, 6 square faces
- 12 vertices
- 24 edges

One copy each of this and the following page are needed to make the polyhedron.

7a. Cuboctahedron with butterfly, beetle, and dragonfly tessellation

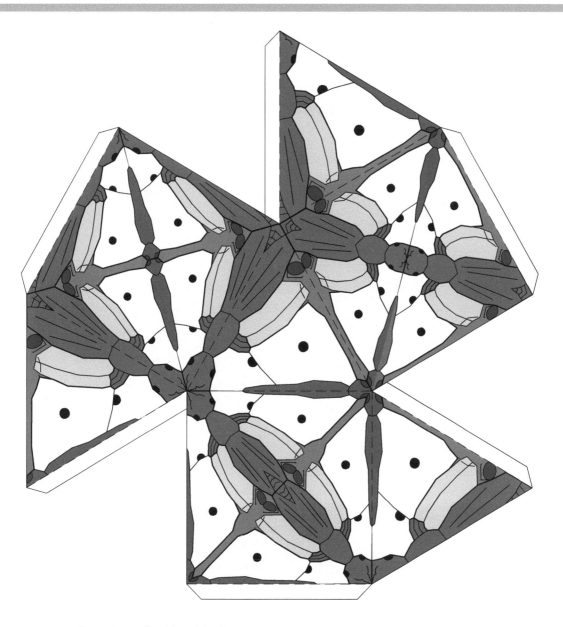

Second page for this polyhedron

7a. Cuboctahedron with butterfly, beetle, and dragonfly tessellation

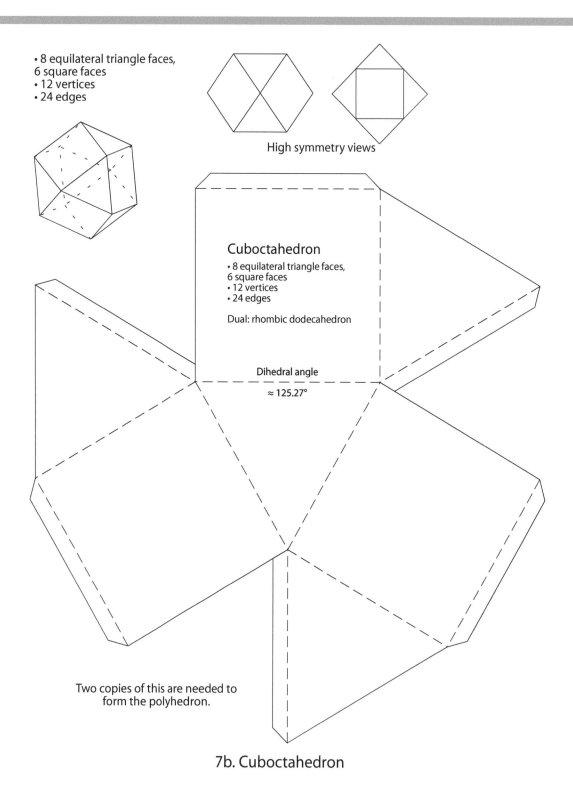

- 8 equilateral triangle faces, 6 square faces
- 12 vertices
- 24 edges

High symmetry views

Cuboctahedron

- 8 equilateral triangle faces, 6 square faces
- 12 vertices
- 24 edges

Dual: rhombic dodecahedron

Dihedral angle

≈ 125.27°

Two copies of this are needed to form the polyhedron.

7b. Cuboctahedron

- 8 hexagon faces, 6 square faces
- 24 vertices
- 36 edges

Other high-symmetry views

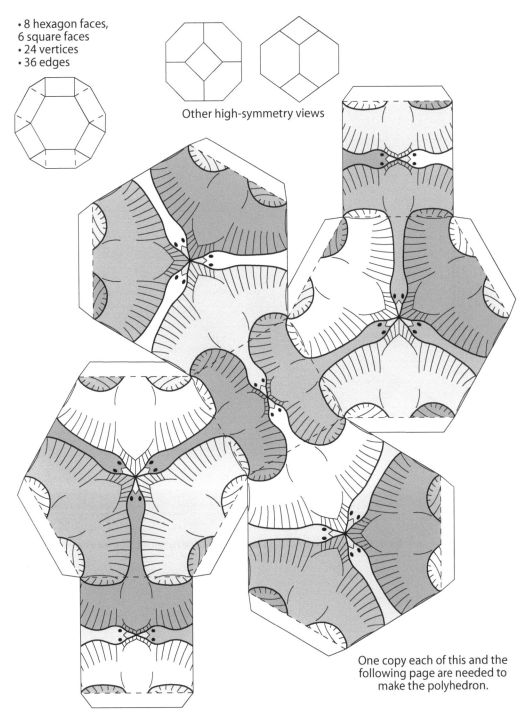

One copy each of this and the following page are needed to make the polyhedron.

8a. Truncated octahedron with goose tessellation

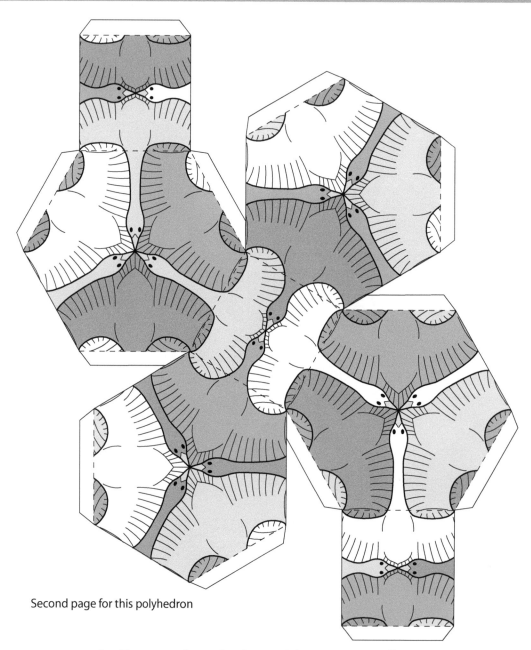

Second page for this polyhedron

8a. Truncated octahedron with goose tessellation

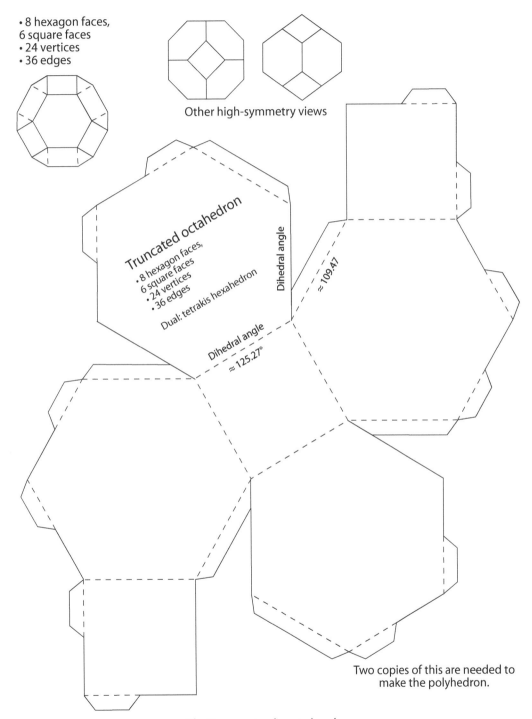

- 8 hexagon faces,
6 square faces
- 24 vertices
- 36 edges

Other high-symmetry views

Truncated octahedron
- 8 hexagon faces,
6 square faces
- 24 vertices
- 36 edges

Dual: tetrakis hexahedron

Dihedral angle ≈ 109.47

Dihedral angle ≈ 125.27°

Two copies of this are needed to make the polyhedron.

8b. Truncated octahedron

- 12 regular faces,
20 equilateral triangle faces
- 30 vertices
- 60 edges

Other high symmetry views.

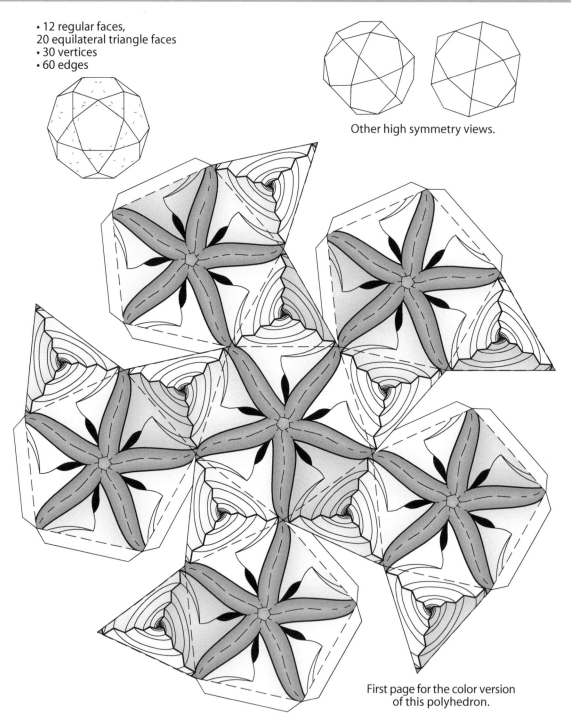

First page for the color version
of this polyhedron.

9a. Icosidodecahedron with starfish and seashell tessellation

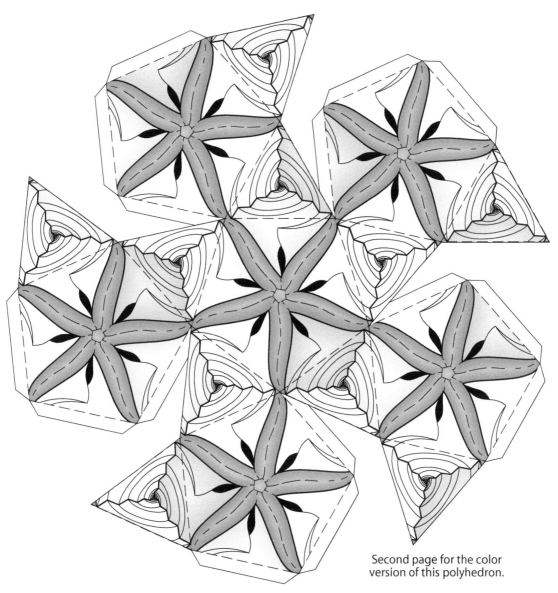

Second page for the color
version of this polyhedron.

9a. Icosidodecahedron with starfish and seashell tessellation

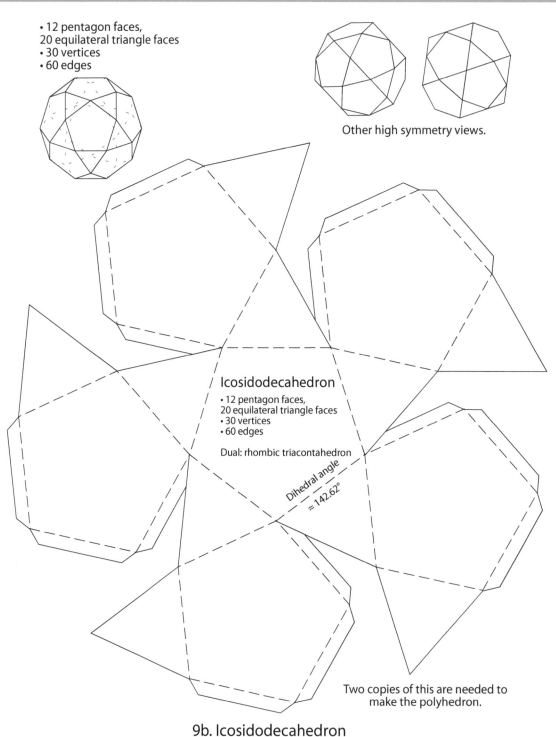

- 12 pentagon faces,
20 equilateral triangle faces
- 30 vertices
- 60 edges

Other high symmetry views.

Icosidodecahedron

- 12 pentagon faces,
20 equilateral triangle faces
- 30 vertices
- 60 edges

Dual: rhombic triacontahedron

Dihedral angle
≈ 142.62°

Two copies of this are needed to
make the polyhedron.

9b. Icosidodecahedron

- 18 square faces,
  8 equilateral triangle faces
- 24 vertices
- 48 edges

Other high symmetry views.

Two copies of this are needed to form the full polyhedron. Note that the top and bottom flowers shown here will double over on each other in the assembled polyhedron. You may want to cut one off of each half.

10a. Small rhombicuboctahedron with flower tessellation

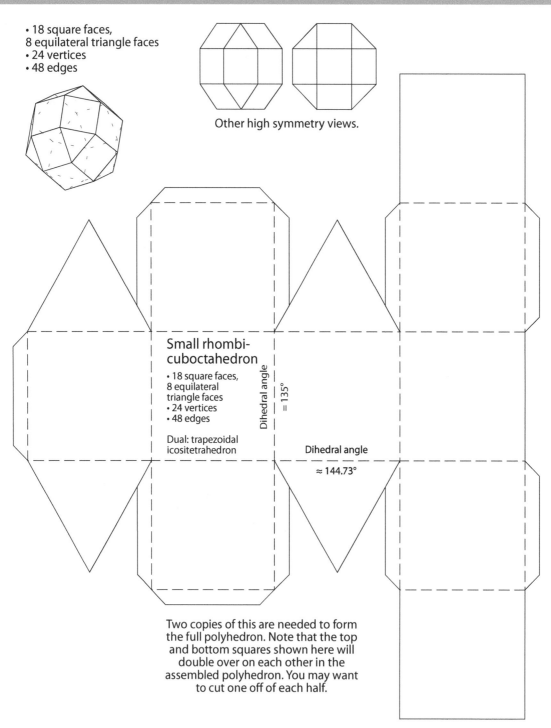

- 18 square faces,
  8 equilateral triangle faces
- 24 vertices
- 48 edges

Other high symmetry views.

Small rhombi-
cuboctahedron

- 18 square faces,
  8 equilateral
  triangle faces
- 24 vertices
- 48 edges

Dual: trapezoidal
icositetrahedron

Dihedral angle = 135°

Dihedral angle
≈ 144.73°

Two copies of this are needed to form
the full polyhedron. Note that the top
and bottom squares shown here will
double over on each other in the
assembled polyhedron. You may want
to cut one off of each half.

10b. Small rhombicuboctahedron

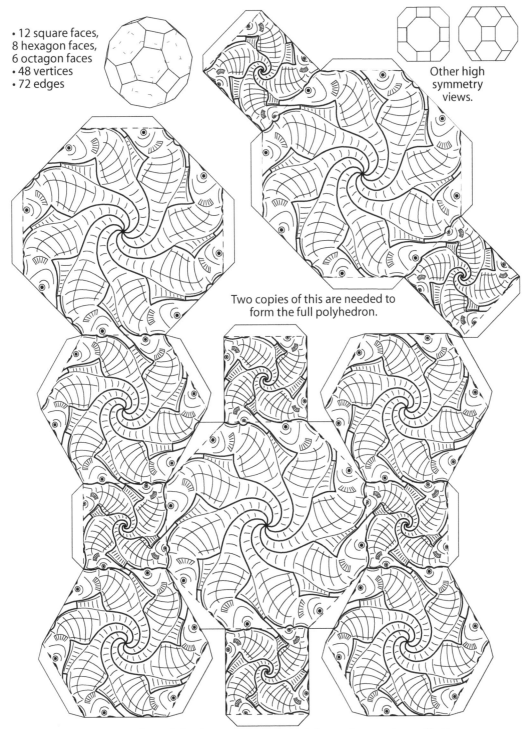

- 12 square faces,
  8 hexagon faces,
  6 octagon faces
- 48 vertices
- 72 edges

Other high
symmetry
views.

Two copies of this are needed to
form the full polyhedron.

11a. Great rhombicuboctahedron with seahorse tessellation

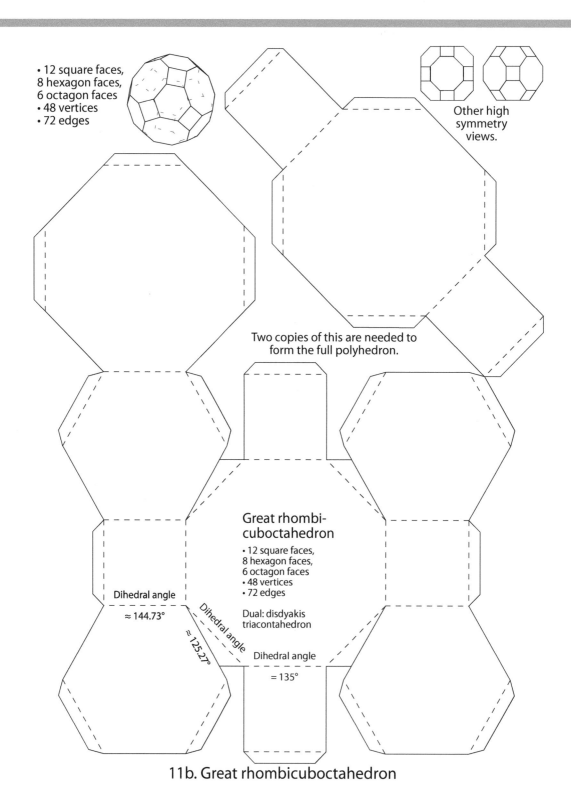

- 12 square faces,
8 hexagon faces,
6 octagon faces
- 48 vertices
- 72 edges

Other high
symmetry
views.

Two copies of this are needed to
form the full polyhedron.

Great rhombi-cuboctahedron

- 12 square faces,
8 hexagon faces,
6 octagon faces
- 48 vertices
- 72 edges

Dual: disdyakis
triacontahedron

Dihedral angle
≈ 144.73°

Dihedral angle
≈ 125.27°

Dihedral angle

Dihedral angle
= 135°

11b. Great rhombicuboctahedron

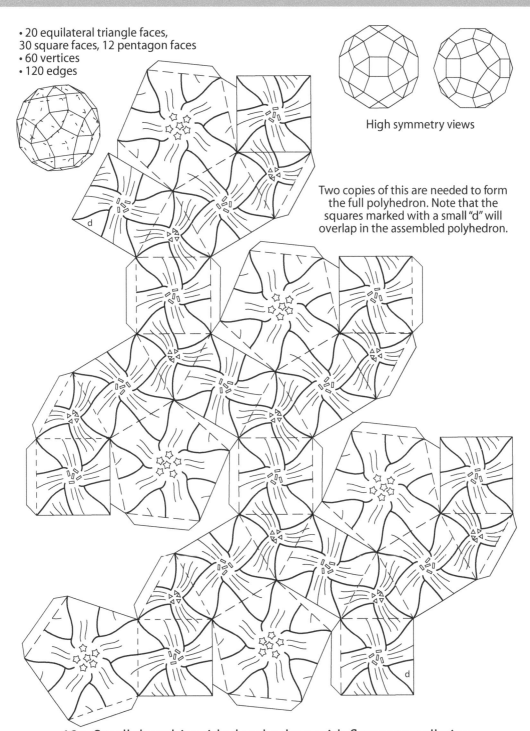

- 20 equilateral triangle faces,
  30 square faces, 12 pentagon faces
- 60 vertices
- 120 edges

High symmetry views

Two copies of this are needed to form the full polyhedron. Note that the squares marked with a small "d" will overlap in the assembled polyhedron.

12a. Small rhombicosidodecahedron with flower tessellation

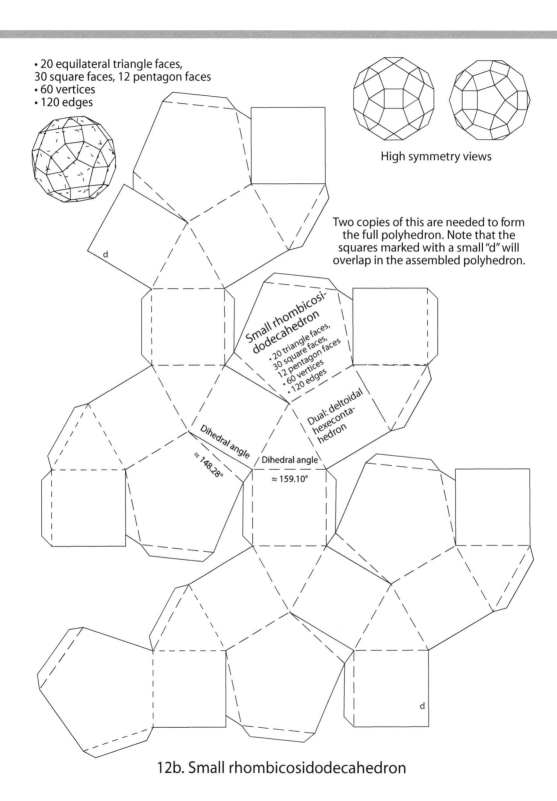

- 20 equilateral triangle faces, 30 square faces, 12 pentagon faces
- 60 vertices
- 120 edges

High symmetry views

Two copies of this are needed to form the full polyhedron. Note that the squares marked with a small "d" will overlap in the assembled polyhedron.

d

Small rhombicosi-dodecahedron

- 20 triangle faces,
30 square faces,
12 pentagon faces
- 60 vertices
- 120 edges

Dual: deltoidal hexeconta-hedron

Dihedral angle
≈ 148.28°

Dihedral angle
≈ 159.10°

d

12b. Small rhombicosidodecahedron

- 12 decagon faces, 20 hexagon faces, and 30 square faces
- 120 vertices
- 180 edges

Three copies of this are needed to form the polyhedron. When assembling the three sections, the hexagons marked "A" should all overlap each other, and similarly for those marked "B". You may wish to cut off two of the three copies of these hexagons once you see how to align them.

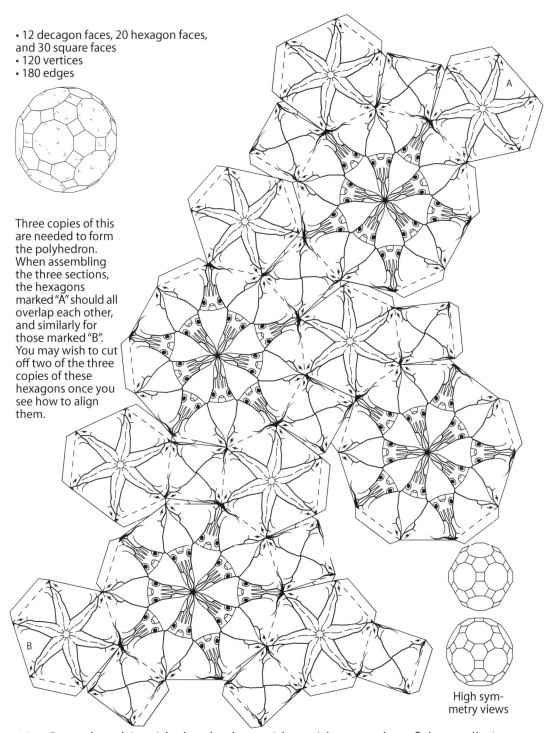

A

B

High symmetry views

13a. Great rhombicosidodecahedron with squid, ray, and starfish tessellation

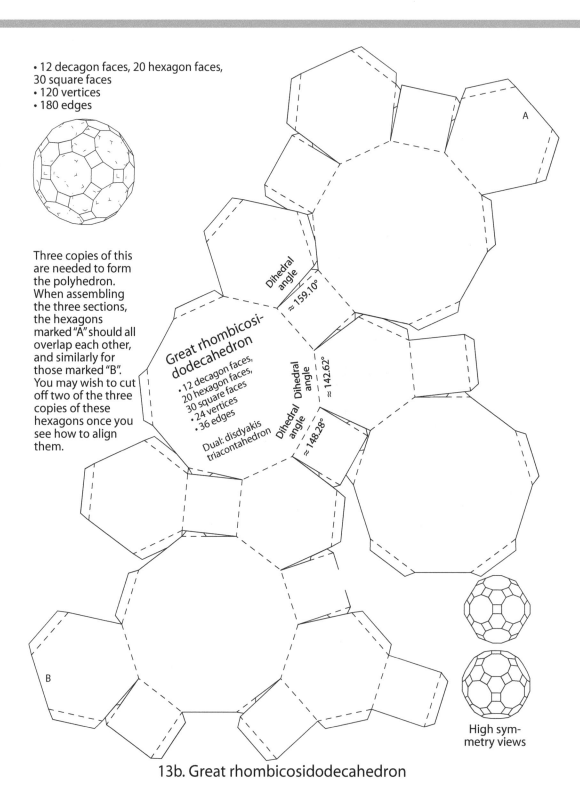

• 12 decagon faces, 20 hexagon faces, 30 square faces
• 120 vertices
• 180 edges

Three copies of this are needed to form the polyhedron. When assembling the three sections, the hexagons marked "A" should all overlap each other, and similarly for those marked "B". You may wish to cut off two of the three copies of these hexagons once you see how to align them.

A

Great rhombicosi-dodecahedron
• 12 decagon faces,
20 hexagon faces,
30 square faces
• 24 vertices
• 36 edges

Dual: disdyakis triacontahedron

Dihedral angle ≈ 159.10°

Dihedral angle ≈ 142.62°

Dihedral angle ≈ 148.28°

B

High symmetry views

13b. Great rhombicosidodecahedron

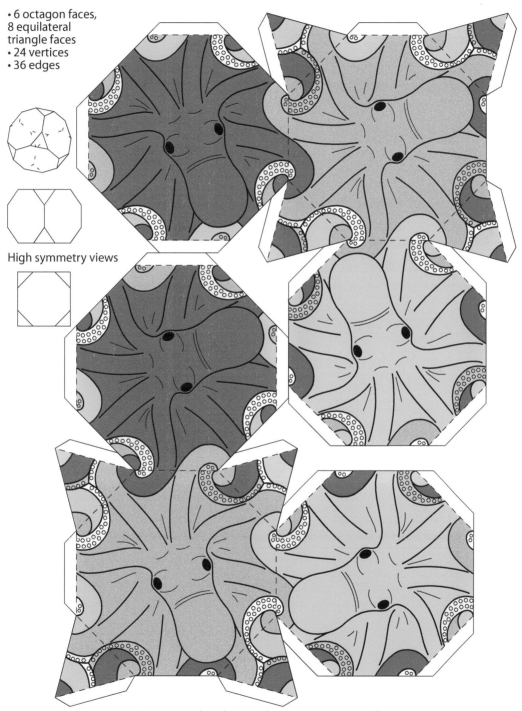

- 6 octagon faces, 8 equilateral triangle faces
- 24 vertices
- 36 edges

High symmetry views

14a. Truncated cube with octopus tessellation

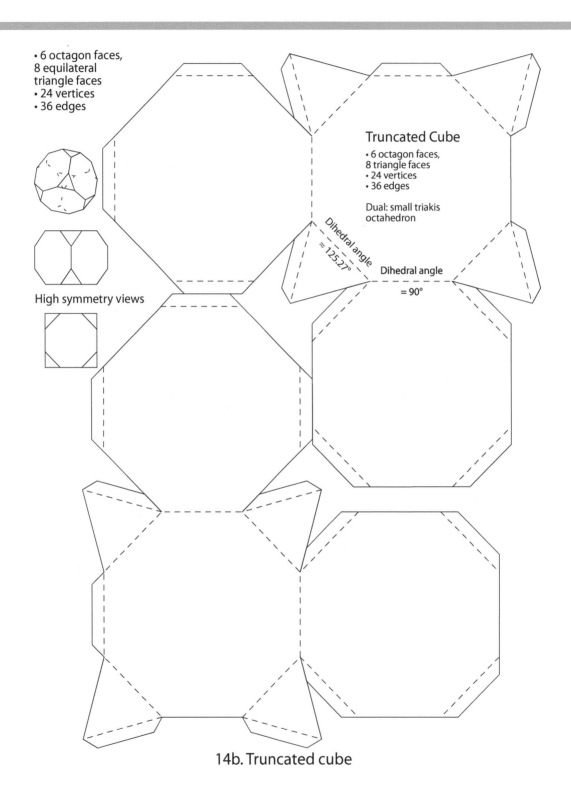

- 6 octagon faces, 8 equilateral triangle faces
- 24 vertices
- 36 edges

High symmetry views

Truncated Cube
- 6 octagon faces, 8 triangle faces
- 24 vertices
- 36 edges

Dual: small triakis octahedron

Dihedral angle ≈ 125.27°

Dihedral angle = 90°

14b. Truncated cube

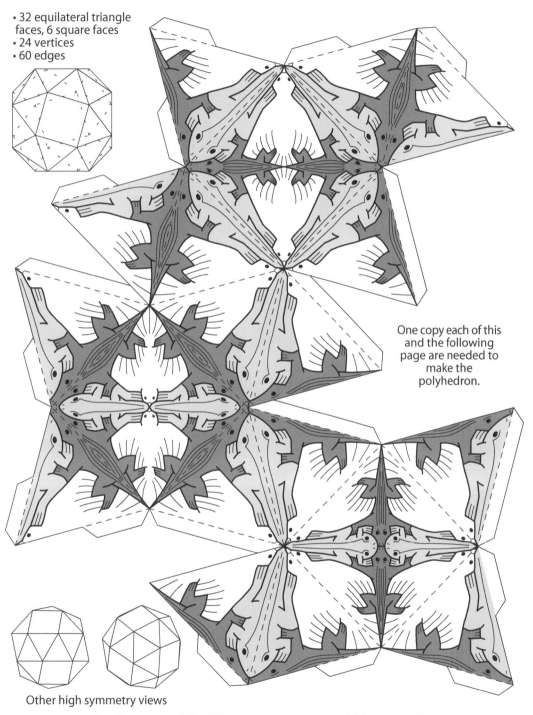

- 32 equilateral triangle faces, 6 square faces
- 24 vertices
- 60 edges

One copy each of this and the following page are needed to make the polyhedron.

Other high symmetry views

15a. Snub cube with alligator, tadpole, and bird tessellation

Second page for
this polyhedron

15a. Snub cube with alligator, tadpole, and bird tessellation

Tessellations: Mathematics, Art, and Recreation

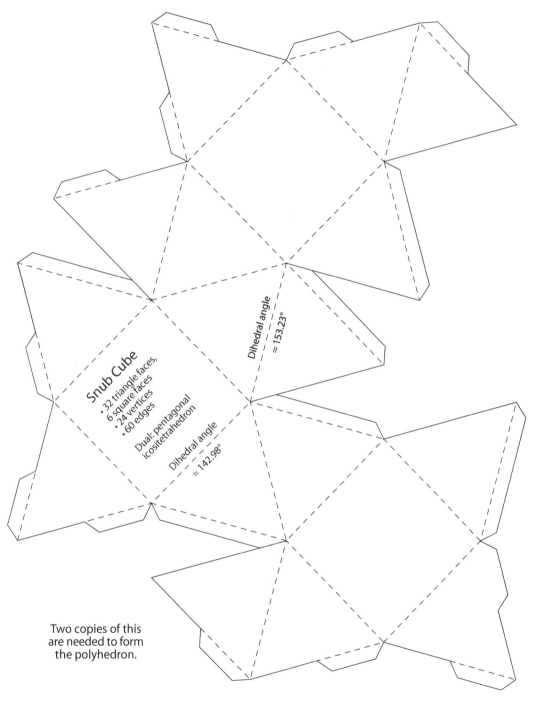

Snub Cube
- 32 triangle faces,
- 6 square faces
- 24 vertices
- 60 edges

Dual: pentagonal
icositetrahedron

Dihedral angle
≈ 153.23°

Dihedral angle
≈ 142.98°

Two copies of this
are needed to form
the polyhedron.

15b. Snub cube

- 12 decagon faces, 20 equilateral triangle faces
- 60 vertices
- 90 edges

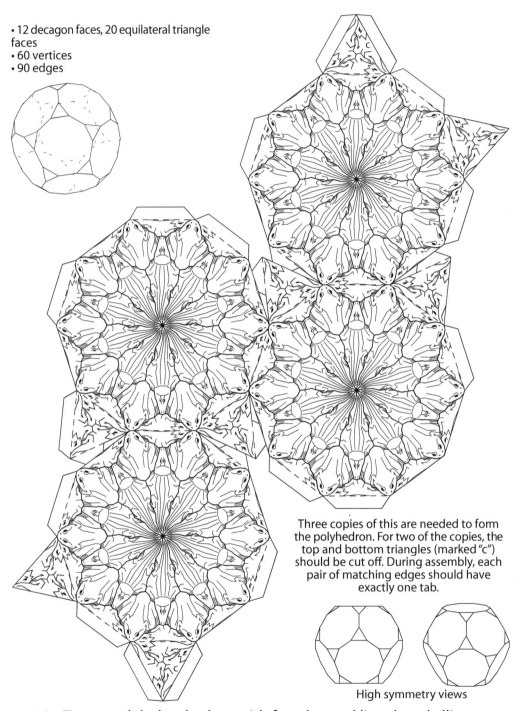

Three copies of this are needed to form the polyhedron. For two of the copies, the top and bottom triangles (marked "c") should be cut off. During assembly, each pair of matching edges should have exactly one tab.

High symmetry views

16a. Truncated dodecahedron with frog, horned lizard, and alligator tessellation

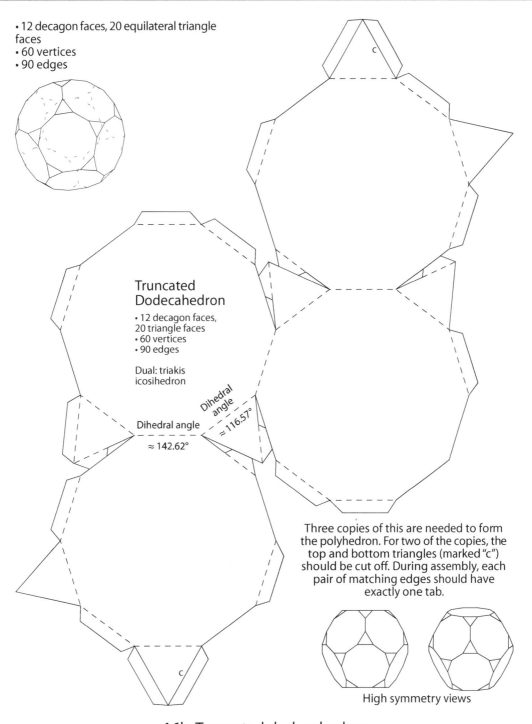

- 12 decagon faces, 20 equilateral triangle faces
- 60 vertices
- 90 edges

Truncated Dodecahedron
- 12 decagon faces, 20 triangle faces
- 60 vertices
- 90 edges

Dual: triakis icosihedron

Dihedral angle ≈ 142.62°

Dihedral angle ≈ 116.57°

Three copies of this are needed to form the polyhedron. For two of the copies, the top and bottom triangles (marked "c") should be cut off. During assembly, each pair of matching edges should have exactly one tab.

High symmetry views

16b. Truncated dodecahedron

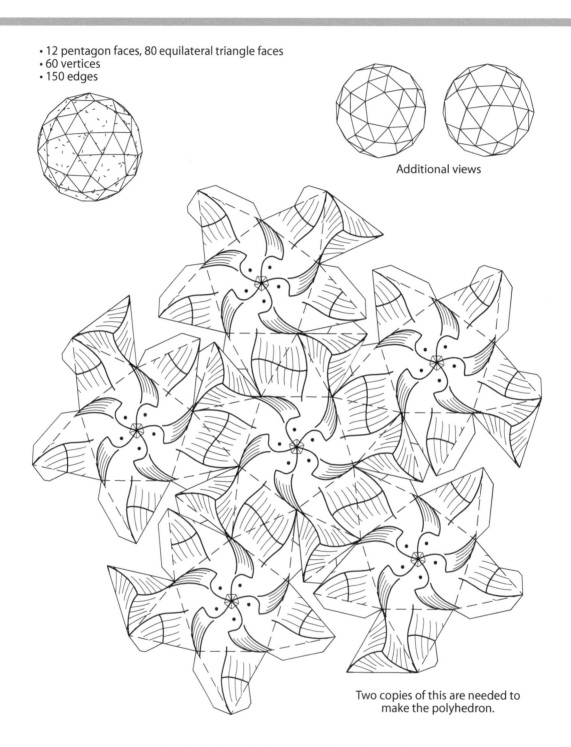

- 12 pentagon faces, 80 equilateral triangle faces
- 60 vertices
- 150 edges

Additional views

Two copies of this are needed to make the polyhedron.

17a. Snub dodecahedron with bird tessellation

- 12 pentagon faces, 80 equilateral triangle faces
- 60 vertices
- 150 edges

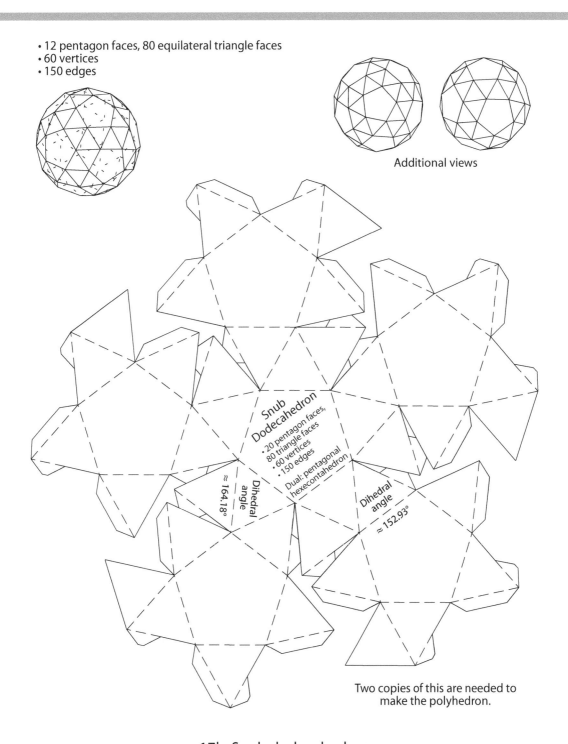

Additional views

Snub
Dodecahedron
- 20 pentagon faces,
  80 triangle faces
- 60 vertices
- 150 edges

Dual: pentagonal
hexecontahedron

Dihedral
angle
≈ 164.18°

Dihedral
angle
≈ 152.93°

Two copies of this are needed to
make the polyhedron.

17b. Snub dodecahedron

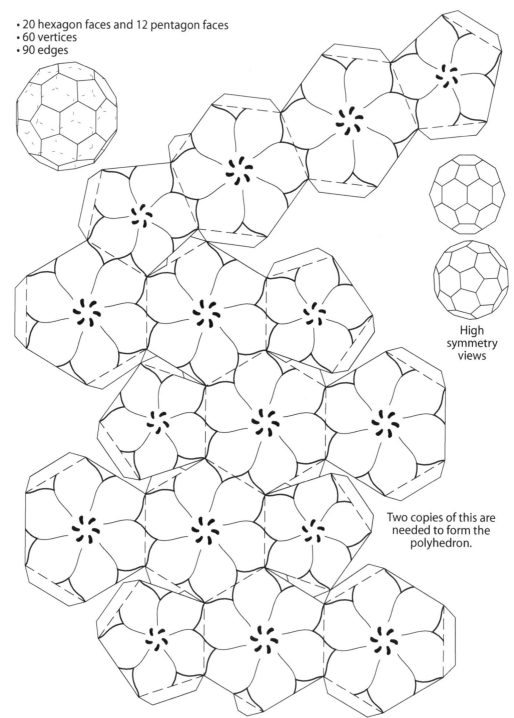

• 20 hexagon faces and 12 pentagon faces
• 60 vertices
• 90 edges

High symmetry views

Two copies of this are needed to form the polyhedron.

18a. Truncated icosahedron with flower tessellation

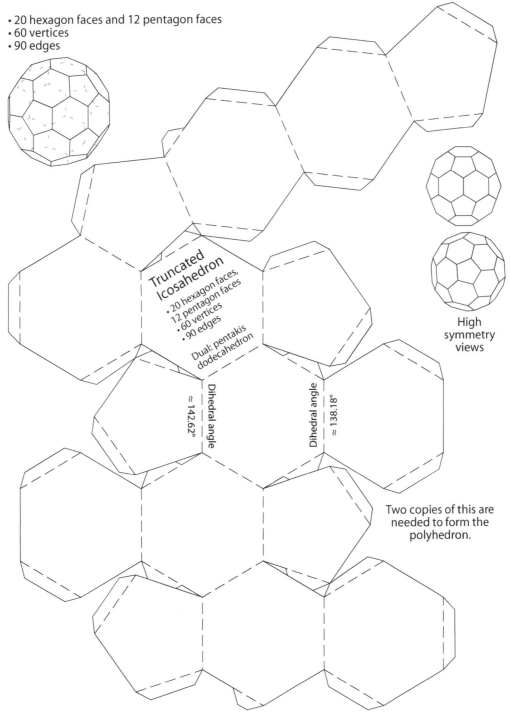

- 20 hexagon faces and 12 pentagon faces
- 60 vertices
- 90 edges

Truncated
Icosahedron

- 20 hexagon faces,
  12 pentagon faces
- 60 vertices
- 90 edges

Dual: pentakis
dodecahedron

Dihedral angle
≈ 142.62°

Dihedral angle
≈ 138.18°

High
symmetry
views

Two copies of this are
needed to form the
polyhedron.

18b. Truncated icosahedron

Tessellation templates for the Archimedean solids

The template in Figure 23.2 is the one used for the paper airplane truncated tetrahedron. Each instance of the template will tile 1/3 of a hexagon and 1/3 of a triangle. For the paper airplanes design, this tile was divided into two smaller tiles. The template in Figure 23.3 can be used for the cuboctahedron. It's not the one used for the butterfly, beetle, and dragonfly cuboctahedron, but is a simpler template. Each instance of this template will tile 1/4 of a square and 1/3 of a triangle. The template in Figure 23.4 can be used for the truncated octahedron. Again, it's a simpler template than the one used for the goose tessellation. This template has two tiles that share a line. Six of

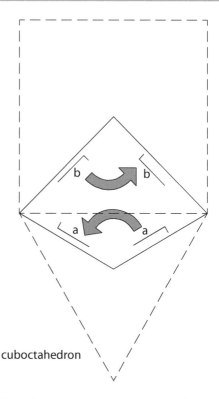

cuboctahedron

Figure 23.3. Template to aid in designing Escheresque tessellations for a cuboctahedron

the large tiles fill a hexagon, while four of the small tiles fill a square.

The template in Figure 23.5 is that used for the seashell and starfish icosidodecahedron. Each instance of the template will tile 1/5 of a pentagon and 1/3 of a triangle. For the seashell and starfish motifs, the template contains one seashell and 1/5 of a starfish. The template in Figure 23.6 is that which was used for the flower small rhombicuboctahedron. There is only one independent line segment here, labeled "a" on some of the segments. Three different tiles are formed by this one segment, as shown. Care should be taken to use the correct square-based tile in each location.

The template in Figure 23.7 is basically the one used for the seahorse great rhombicuboctahedron. Four independent line

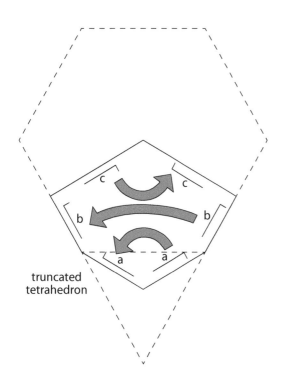

truncated
tetrahedron

Figure 23.2. Template to aid in designing Escheresque tessellations for a truncated tetrahedron.

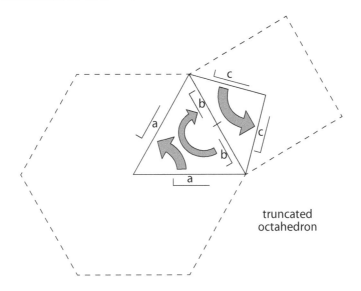

Figure 23.4. Template to aid in designing Escheresque tessellations for a truncated octahedron.

segments are used to form triangles that fill the three different polygons. The template in Figure 23.8, used for the flower small rhombicosidodecahedron, contains two independent line segments.

The template in Figure 23.9 is similar to that which was used for the squid, ray, and starfish great rhombicosidodecahedron, with each of the three different polygons broken down into triangles. It has four independent line segments. The template in Figure 23.10 is the one used for the octopus truncated cube, with two independent line segments. Breaking down the octagon into smaller tiles

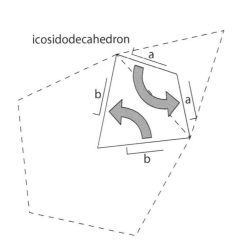

Figure 23.5. Template to aid in designing Escheresque tessellations for a icosidodecahedron.

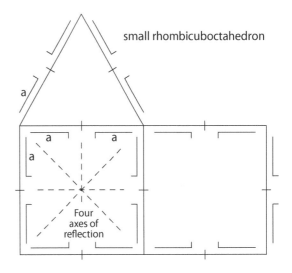

Figure 23.6. Template to aid in designing Escheresque tessellations for a small rhombicuboctahedron.

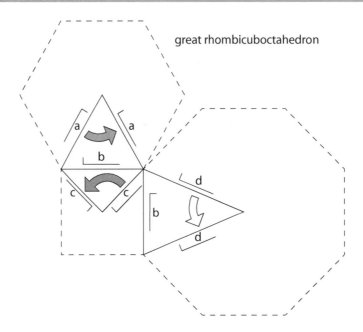

great rhombicuboctahedron

Figure 23.7. Template to aid in designing Escheresque tessellations for a great rhombicuboctahedron.

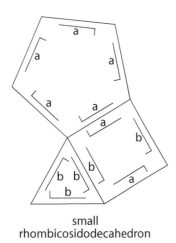

small
rhombicosidodecahedron

Figure 23.8. Template to aid in designing Escheresque tessellations for a small rhombicosidodecahedron.

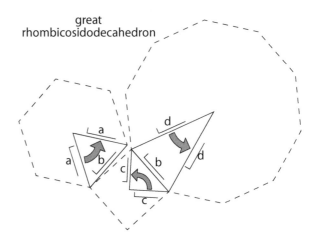

great
rhombicosidodecahedron

Figure 23.9. Template to aid in designing Escheresque tessellations for a great rhombicosidodecahedron.

would allow a wider variety of motifs to be adapted to this polyhedron.

The template in Figure 23.11, with three independent line segments, can be used for the snub cube. It is a simpler template than the one used for the alligator, tadpole, and bird snub cube. The same equilateral triangle template can be used for all triangles in the net. The template in Figure 23.12 is similar to that used for the frog, horned lizard, and alligator truncated dodecahedron, with three independent line segments.

The template in Figure 23.13 is that which was used for the bird snub dodecahedron.

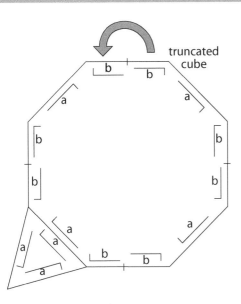

Figure 23.10. Template to aid in designing Escheresque tessellations for a truncated cube.

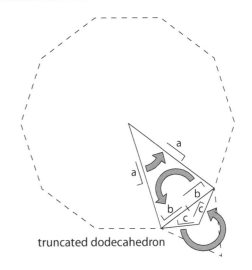

truncated dodecahedron

Figure 23.12. Template to aid in designing Escheresque tessellations for a truncated dodecahedron.

snub dodecahedron

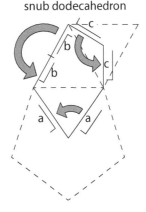

Figure 23.13. Template to aid in designing Escheresque tessellations for a snub dodecahedron.

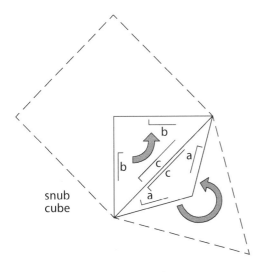

snub cube

Figure 23.11. Template to aid in designing Escheresque tessellations for a snub cube.

The template contains three independent line segments, with the (c) line segments occurring within the triangles that are adjacent only to other triangles. The other type of triangle is adjacent to both pentagons and triangles. Figure 23.14 shows how to apply this template to the polyhedron. The template in Figure 23.15 is that which was used for the flower truncated icosahedron. A single independent line segment is used repeatedly.

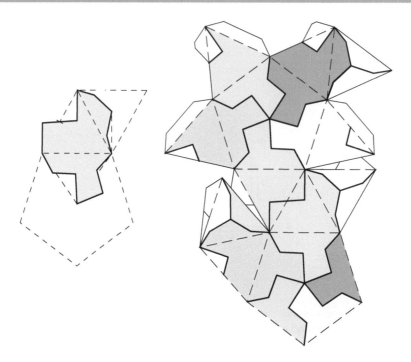

Figure 23.14. Example illustrating how to apply the template of Figure 23.13 to the snub dodecahedron net.

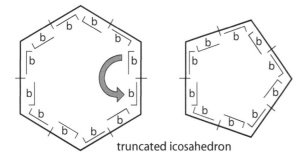

truncated icosahedron

Figure 23.15. Template to aid in designing Escheresque tessellations for a truncated icosahedron.

Activity 23.1. Surface area of Archimedean solids

Materials: Copies of Worksheet 23.1.

Objective: Learn to calculate the surface areas of polyhedra by calculating the areas of the individual faces and summing over all the faces.

Vocabulary: Archimedean solid, face, vertex, edge, cuboctahedron, truncated octahedron, truncated cube.

Specific Common Core State Standard for Mathematics addressed by the activity

Solve real-world and mathematical problems involving area, volume, and surface area of two- and three-dimensional objects composed of triangles, quadrilaterals, polygons, cubes, and right prisms. (7.G 6)

Activity Sequence

1. Pass out the worksheet.
2. Write the vocabulary terms on the board and discuss the meaning of each one.
3. Have the students perform the first task on the worksheet. This requires using the formula for the area of a triangle in terms of its base and height and then adding the areas of the individual faces. Ask a student to share his or her result and to describe how it was calculated.

    $6 + 2\sqrt{3}$.

4. Have the students perform the second task on the worksheet. This involves calculating the area of a regular hexagon. The easiest way to do this is to recognize that the hexagon can be divided into six equilateral triangles. Ask a student to share his or her result and to describe how it was calculated.

    $6 + 12\sqrt{3}$.

5. Have the students use their calculators to get an approximate ratio of the area of these two polyhedra.

    A truncated octahedron has a surface area approximately 2.83 times that of a cuboctahedron with the same edge lengths.

6. Have the students perform the third task on the worksheet. This involves calculating the area of a regular octagon. Ask a student to share his or her result and to describe how it was calculated.

    A regular octagon with unit edge length has an area of $2 + 2\sqrt{2}$. The total surface area of the truncated cube is $12 + 12\sqrt{2} + 2\sqrt{3}$. The cube has surface area $6(1 + \sqrt{2})^2 = 18 + 12\sqrt{2}$. The ratio of the surface area of the truncated cube to that of the cube is approximately 0.927.

Worksheet 23.1. Surface area of Archimedean solids

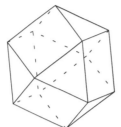
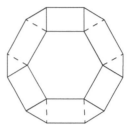

Cuboctahedron   Truncated octahedron

Calculate the surface area of a cuboctahedron with edges of length 1. The height of an equilateral triangle with base equal to 1 is $\sqrt{3}/2$.

Calculate the surface area of a truncated octahedron with edges of length 1. How do the surface areas of these two polyhedra compare?

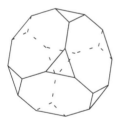
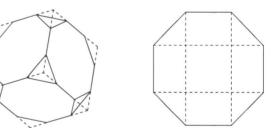

Truncated cube

Calculate the surface area of a truncated cube with edges of length 1. The diagram on the right will help you calculate the area of a regular octagon. If the truncated cube was formed by removing the corners of a cube as shown, how do the surface areas of the two solids compare?

Activity 23.2. Volume of a truncated cube

Materials: Copies of Worksheet 23.2.

Objective: Learn how to use the Pythagorean Theorem to solve a non-trivial problem – calculating the volume of an Archimedean solid.

Vocabulary: Truncated cube.

Specific Common Core State Standard for Mathematics addressed by the activity

Apply the Pythagorean Theorem to determine unknown side lengths in right triangles in real-world and mathematical problems in two and three dimensions. (8.G 7)

Activity Sequence

1. Pass out the worksheet.
2. Write the vocabulary term on the board and discuss its meaning.
     Drawings and properties of these Archimedean solids can be found at the beginning of the chapter.
3. Have the students perform the first task on the worksheet. Ask a student to share his or her result and to describe how it was calculated.
     $1 + \sqrt{2}, (1 + \sqrt{2})3 = 7 + 5\sqrt{2} \approx 14.07.$
4. Have the students perform the second task on the worksheet. Ask a student to share his or her result and to describe how it was calculated.
     $h_T = \sqrt{3}/2.$ *Base area* $= \sqrt{3}/4.$
5. Have the students perform the third task on the worksheet. This is done by recognizing that $d1/(1/2) = (1/2)/h_T$. Ask a student to share his or her result and to describe how it was calculated.
     $d_1 = 1/2\sqrt{3}.$
6. Have the students perform the fourth task on the worksheet. Ask a student to share his or her result and to describe how it was calculated.
     Since d_2 divide a right isosceles triangle with hypotenuse 1 into two half-sized right isosceles triangles, d_2 = 1/2.
7. Have the students perform the fifth task on the worksheet. Ask a student to share his or her result and to describe how it was calculated.
     $h_P = 1/\sqrt{6}. V_P = 1/(12\sqrt{2})$
8. Have the students perform the sixth task on the worksheet. Ask a student to share his or her result and to describe how it was calculated.
     $7 + 5\sqrt{2} - 2/(3\sqrt{2}) \approx 13.60.$ *The ratio of the volume of the truncated cube to the cube is approximately 0.967.*

Worksheet 23.2. Volume of a truncated cube

The goal of this worksheet is to use the Pythagorean Theorem to calculate the volume of a truncated cube with edge lengths equal to 1. The strategy is to calculate the volume of the enclosing cube and the triangular pyramids that need to be removed from the cube to form the truncated cube and then take the difference between them.

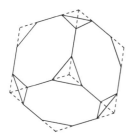 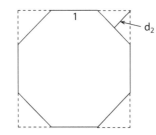 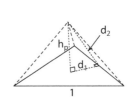 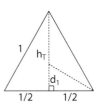

What are the edge length and volume of the enclosing cube?

The volume of each triangular pyramid can be calculated from the formula for the volume of any pyramid, 1/3 (base × height). First, calculate the area of the base by using the Pythagorean Theorem to calculate the height h_T of the equilateral triangle.

The height h_P of the pyramid can be calculated if d_1 and d_2 are known. First calculate d_1 by comparing similar triangles in the fourth diagram in the figure.

Using the second diagram, calculate d_2.

Use d_1 and d_2 to calculate h_P and the volume of one triangular pyramid.

Use the volumes of the cube and pyramid to calculate the volume of the truncated cube. Compare it to that of the cube.

Chapter 24

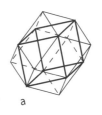

a b c

Tessellating Other Polyhedra

Other popular polyhedra

This chapter provides tessellations and nets for a small sampling of other polyhedra that are frequently encountered. The first polyhedron presented is a 14-gon prism. A (regular right) prism has two congruent regular polygon faces separated by a band of squares. In the 14-gon prism in this chapter, a band of rectangles is used rather than a band of squares in order to better fit the beetle, moth, and bee tessellation. The second polyhedron in this chapter is a hexagonal antiprism. In an antiprism, the regular polygons are rotated with respect to one another so that a band of triangles (equilateral if the antiprism is regular) can connect them.

The third polyhedron in this chapter is a heptagonal pyramid. A (regular right) pyramid consists of a regular polygon (the "base") and triangles meeting at a common vertex (the "apex") located at a point that lies along a line including the centroid of the base and perpendicular to the base. A dipyramid is formed by two pyramids joined base-to-base.

The fourth polyhedron in this chapter is a rhombic dodecahedron, a solid with 12 faces that are rhombi. The rhombic dodecahedron has a variety of interesting properties. It is the dual of the cuboctahedron. It is also a space-filling polyhedron, meaning it tiles in three dimensions. Some minerals, for example garnet, form rhombic dodecahedral crystals naturally. A rhombic dodecahedron

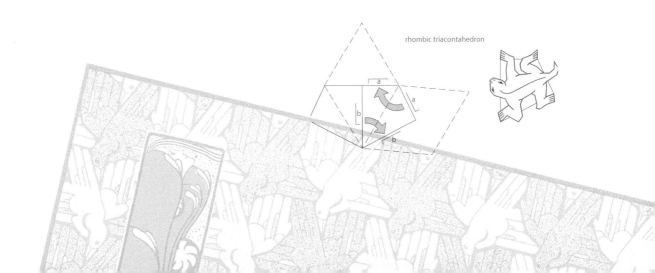

rhombic triacontahedron

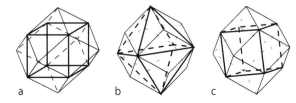

a   b   c

Figure 24.1. Relationships of the rhombic dodecahedron to other polyhedra. (a) Forming a rhombic
dodecahedron by placing square pyramids on each
face of a cube. (b) The long diagonals of the faces are
the edges of an octahedron. (c) The short diagonals of
the faces are the edges of a cube.

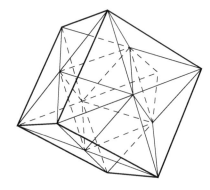

Figure 24.3. The stella octangula fits neatly inside a
cube.

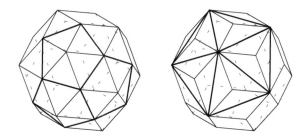

Figure 24.2. The short diagonals of a rhombic triacontahedron form the edges of a dodecahedron, while the
long diagonals form the edges of an icosahedron.

can be constructed by placing a square
pyramid of height 1/2 on each face of a unit
cube, as shown in Figure 24.1a. An octahedron is obtained from the long diagonals
of a rhombic dodecahedron (Figure 24.1b),
while a cube is obtained from the short
diagonals (Figure 24.1c). The ratio of the

long to short diagonals in each rhombus is
the square root of 2.

The next to last polyhedron in this chapter
is a rhombic triacontahedron. It has 30 faces,
each of which is a Golden rhombus: one for
which the ratio of the long to short diagonals
is in the golden mean, 1.618 … . It is the dual
of the icosidodecahedron. A dodecahedron is
obtained from the short diagonals of a rhombic triacontahedron, while an icosahedron is
obtained from the long diagonals, as shown in
Figure 24.2 (three faces of the dodecahedron
are shown and five faces of the icosahedron).

The last polyhedron in this chapter is the
stella octangula. This can be seen as two
interpenetrating tetrahedra. It is the only stellation of the octahedron and fits neatly inside
a cube, as shown in Figure 24.3. The name
was given by Johannes Kepler.

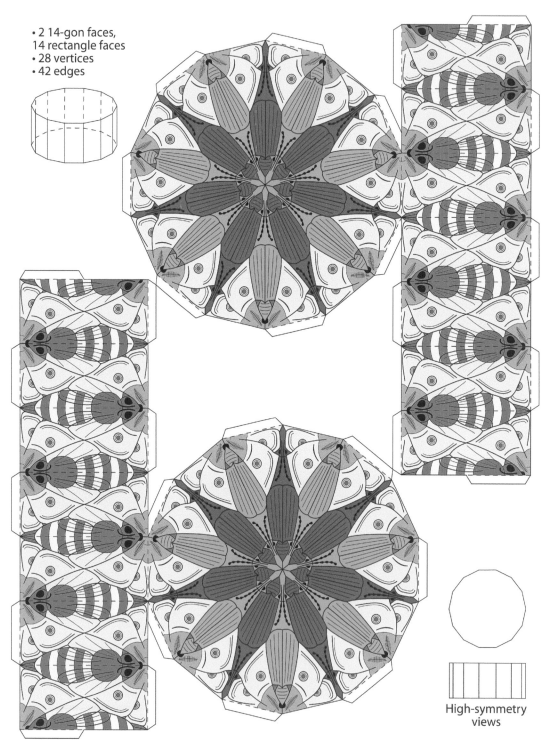

- 2 14-gon faces,
  14 rectangle faces
- 28 vertices
- 42 edges

High-symmetry views

19a. 14-gon prism with beetles, moths, and bees tessellation

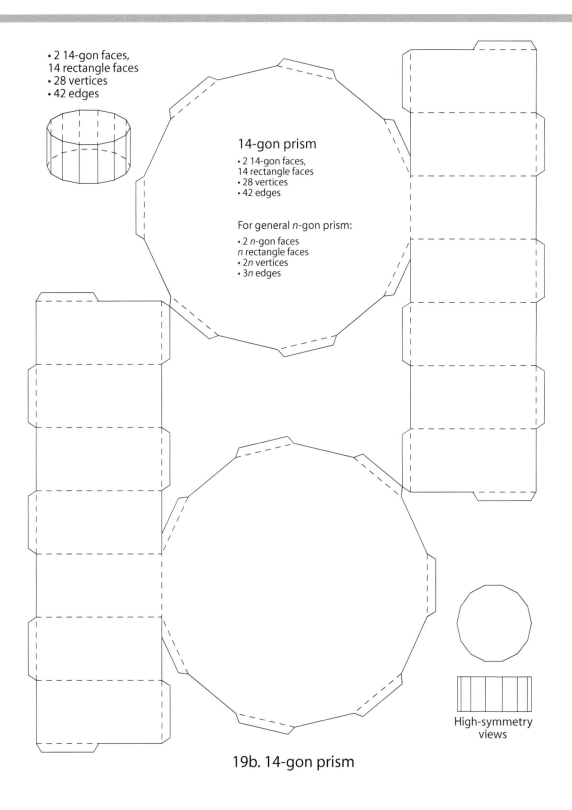

- 2 14-gon faces,
14 rectangle faces
- 28 vertices
- 42 edges

14-gon prism

- 2 14-gon faces,
14 rectangle faces
- 28 vertices
- 42 edges

For general *n*-gon prism:

- 2 *n*-gon faces
n rectangle faces
- 2*n* vertices
- 3*n* edges

High-symmetry views

19b. 14-gon prism

- 2 hexagon faces,
12 isosceles triangle faces (not necessarily equilateral)
- 12 vertices
- 24 edges

Two copies of this are needed to
form the polyhedron.

High-symmetry
views

20a. Hexagonal antiprism with snowflake tessellation

- 2 hexagon faces,
12 isosceles triangle faces (not necessarily equilateral)
- 12 vertices
- 24 edges

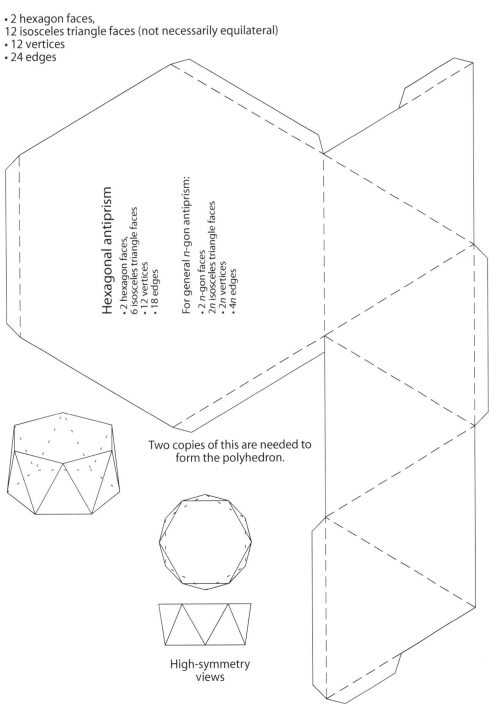

Hexagonal antiprism

- 2 hexagon faces,
6 isosceles triangle faces
- 12 vertices
- 18 edges

For general *n*-gon antiprism:
- 2 *n*-gon faces
2*n* isosceles triangle faces
- 2*n* vertices
- 4*n* edges

Two copies of this are needed to
form the polyhedron.

High-symmetry
views

20b. Hexagonal antiprism

- 1 heptagon face,
  7 isosceles triangle faces
- 8 vertices
- 14 edges

High-symmetry views

21a. Heptagonal pyramid with tyrannosaurus, pteranodon, and euoplocephalus tessellation

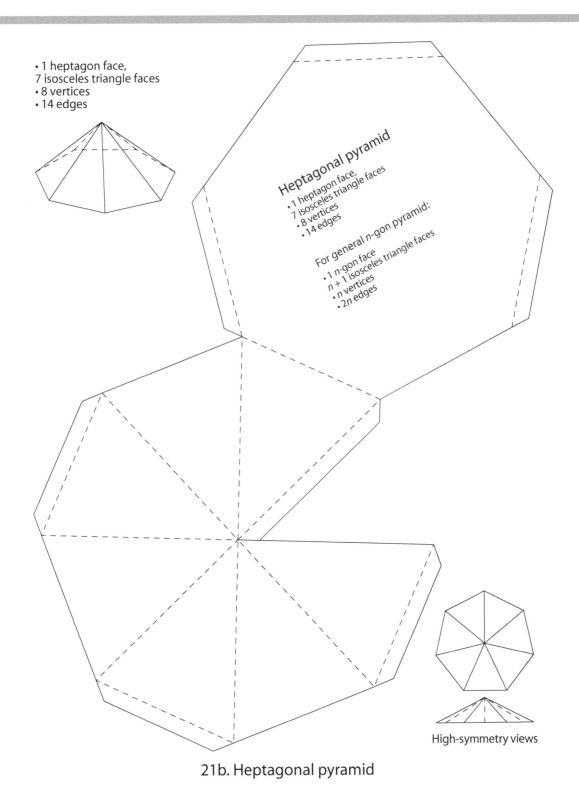

- 1 heptagon face,
7 isosceles triangle faces
- 8 vertices
- 14 edges

Heptagonal pyramid

- 1 heptagon face,
7 isosceles triangle faces
- 8 vertices
- 14 edges

For general *n*-gon pyramid:

- 1 *n*-gon face
n + 1 isosceles triangle faces
- *n* vertices
- 2*n* edges

High-symmetry views

21b. Heptagonal pyramid

- 12 rhombus faces
- 14 vertices
- 24 edges

Three copies of this are needed to form the full polyhedron.

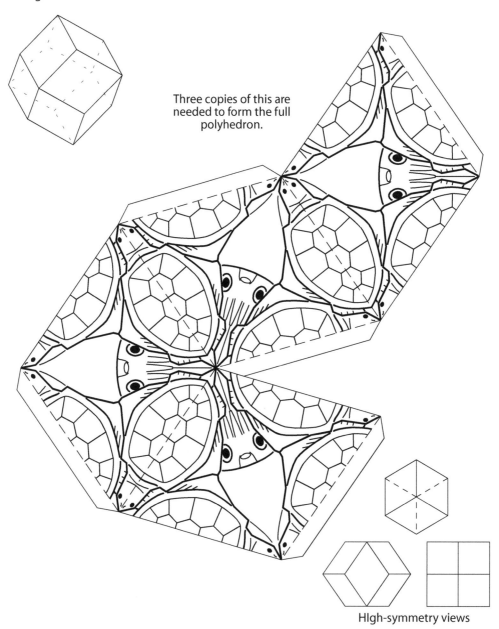

High-symmetry views

22a. Rhombic dodecahedron with squid and sea turtle tessellation

- 12 rhombus faces
- 14 vertices
- 24 edges

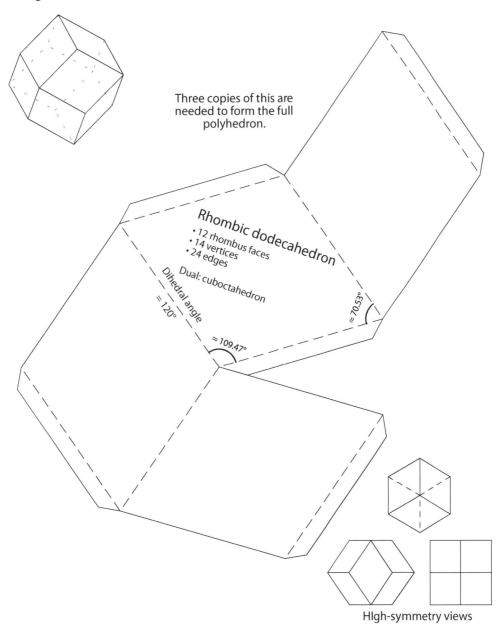

Three copies of this are needed to form the full polyhedron.

Rhombic dodecahedron
- 12 rhombus faces
- 14 vertices
- 24 edges

Dual: cuboctahedron

Dihedral angle
≈ 120°

≈ 109.47°

≈ 70.53°

High-symmetry views

22b. Rhombic dodecahedron

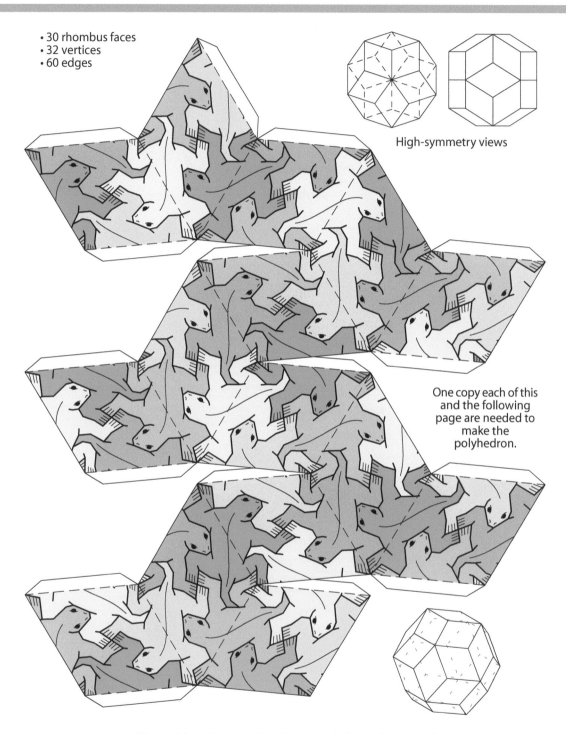

- 30 rhombus faces
- 32 vertices
- 60 edges

High-symmetry views

One copy each of this and the following page are needed to make the polyhedron.

23a. Rhombic triacontahedron with lizard tesssellation

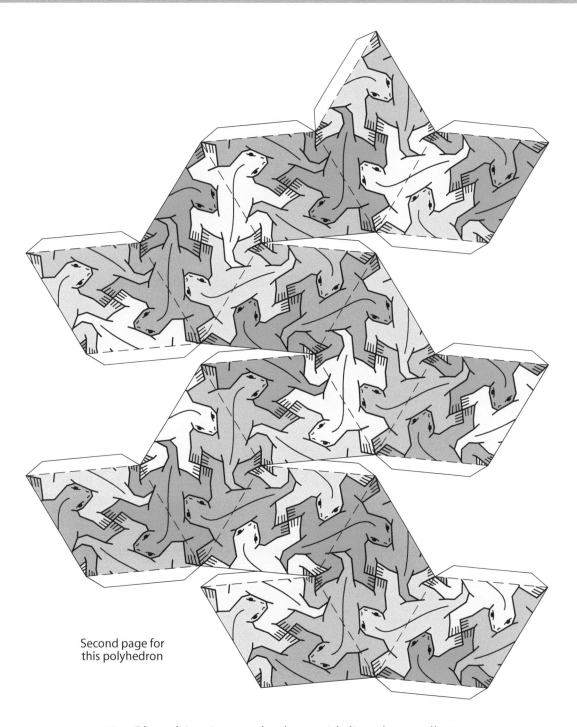

Second page for
this polyhedron

23a. Rhombic triacontahedron with lizard tesssellation

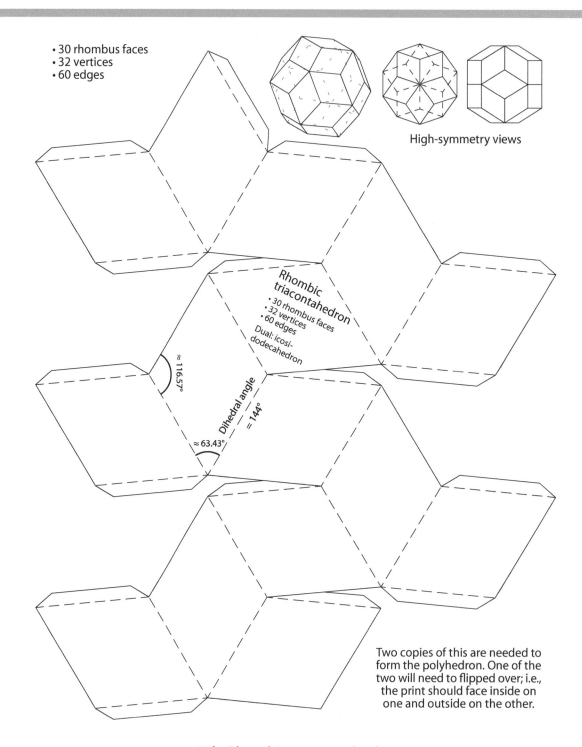

- 30 rhombus faces
- 32 vertices
- 60 edges

High-symmetry views

Rhombic
triacontahedron
- 30 rhombus faces
- 32 vertices
- 60 edges
Dual: icosi-
dodecahedron

≈ 116.57°

Dihedral angle
≈ 144°

≈ 63.43°

Two copies of this are needed to
form the polyhedron. One of the
two will need to flipped over; i.e.,
the print should face inside on
one and outside on the other.

23b. Rhombic triacontahedron

- 24 equilateral triangle faces
- 14 vertices
- 36 edges

Two copies of this are needed to form the polyhedron.

The evenly dashed lines indicate mountain folds, while the shot-long dashed lines indicate valley folds.

High symmetry views

24a. Stella octangula with gorilla tessellation

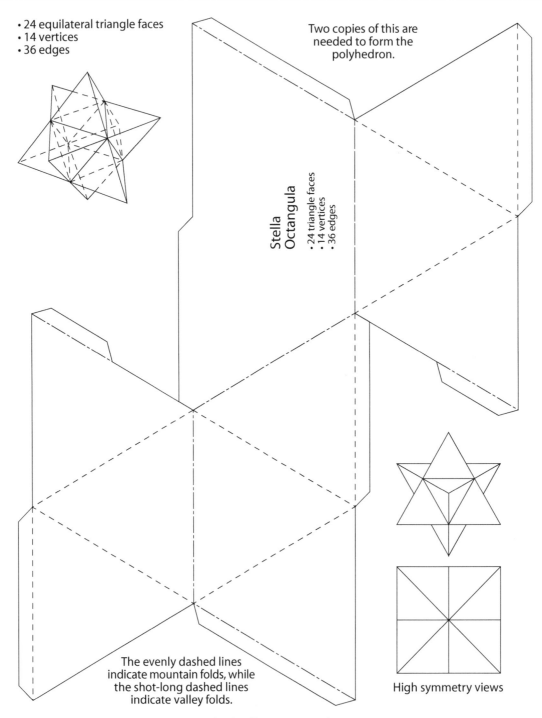

- 24 equilateral triangle faces
- 14 vertices
- 36 edges

Two copies of this are needed to form the polyhedron.

Stella Octangula
- 24 triangle faces
- 14 vertices
- 36 edges

The evenly dashed lines indicate mountain folds, while the shot-long dashed lines indicate valley folds.

High symmetry views

24b. Stella octangula

Tessellation templates

The template in Figure 24.4 for the 14-gon prism is simpler than that used for the beetle, moth, and bee tessellation. It involves three independent line segments. Note that the two instances of the b line segment are rotated to 180° with respect to one another. The template provided for the hexagonal antiprism (Figure 24.5) is also simpler than that used for the snowflake tessellation. It only has a single independent line segment, which is used twice to form each edge of the polyhedron, with the two uses related by a 180° rotation. The hexagon can be tiled with six of the equilateral triangles, or it can be a different tile or tiles, such as a single hexagon or three rhombi.

The template in Figure 24.6 for the heptagonal pyramid is simpler than that used for the dinosaur tessellation. It involves three independent line segments. Note

that (c) can be an elongated version of (a), so that the same motif can be used for the two triangles, just stretched a bit for the triangular faces. The template provided for the rhombic dodecahedron (Figure 24.7) is the same as that used for the squid and sea turtle tessellation. It results in two different tiles with bilateral symmetry. This is a convenient symmetry for creatures seen from the front or top, since most animals and insects possess approximate overall bilateral symmetry.

The template in Figure 24.8 for the rhombic triacontahedron is that which was used for the lizard tessellation. It has two independent line segments that form a kite-shaped tile. The lizard motif doesn't look at all like the kite, so the relationship between the two is shown. The template provided for the stella octangula (Figure 24.9) is also that which was used for the gorilla tessellation. It's similar to the template for the rhombic triacontahedron,

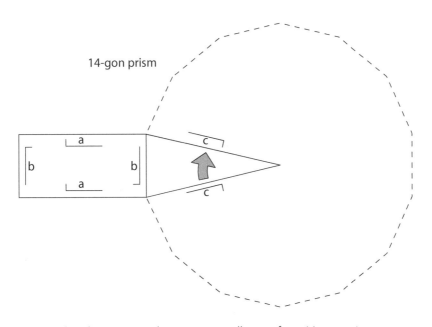

Figure 24.4. Template to aid in designing Escheresque tessellations for a 14-gon prism.

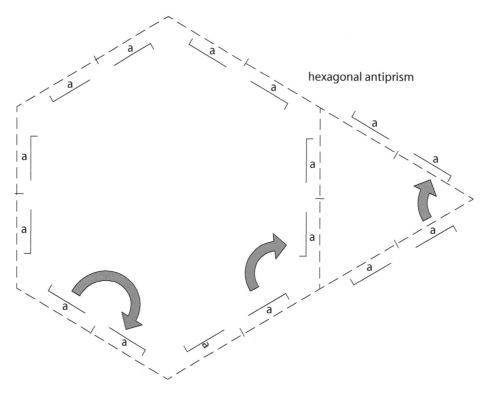

Figure 24.5. Template to aid in designing Escheresque tessellations for a hexagonal antiprism.

though the angles are not quite the same. Again, the relationship of the gorilla to the kite is shown. You'll notice there is some similarity in these two tessellating motifs.

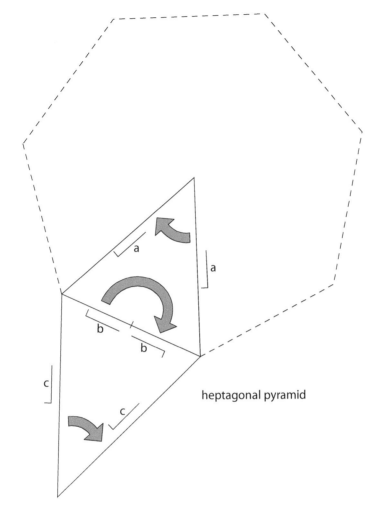

Figure 24.6. Template to aid in designing Escheresque tessellations for a heptagonal pyramid.

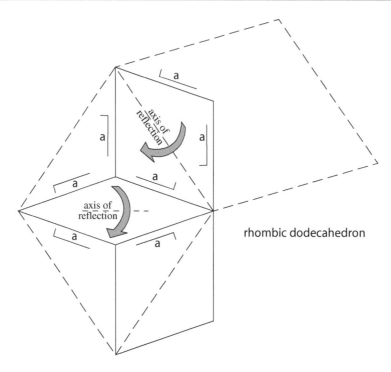

Figure 24.7. Template to aid in designing Escheresque tessellations for a rhombic dodecahedron.

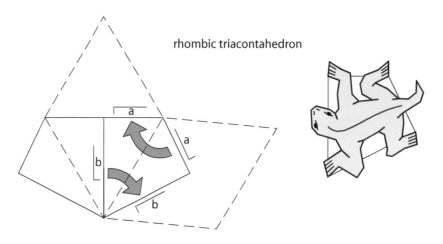

Figure 24.8. Template to aid in designing Escheresque tessellations for a rhombic triacontahedron.

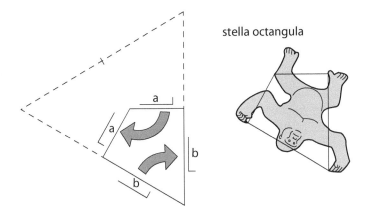

Figure 24.9. Template to aid in designing Escheresque tessellations for a stella octangula.

Activity 24.1. Cross-sections of polyhedra

Materials: Copies of Worksheet 24.1 for students.

Objective: Learn to visualize, describe, and draw polygons that result from slicing polyhedra with planes.

Vocabulary: Pyramid, dipyramid, prism, antiprism.

Specific Common Core State Standard for Mathematics addressed by the activity

Describe the two-dimensional figures that result from slicing three-dimensional figures, as in plane sections of right rectangular prisms and right rectangular pyramids. (7.G3)

Activity Sequence

1. Pass out the worksheet.
2. Write the vocabulary terms on the board and discuss the meaning of each one.
3. Have the students draw the cross-sections for the right square pyramid. Have a student share his or her result for each drawing. The first drawing should be an isosceles triangle, and the second drawing should be a trapezoid. The base of the isosceles triangle has length 1.

You may want to have the students calculate the lengths of the other two sides.

4. Have the students draw the cross-sections for the triangular dipyramid. Have a student share his or her result for each drawing. The first drawing should be an equilateral triangle, and the second drawing should be a rhombus.
5. Have the students draw the cross-sections for the hexagonal prism. Have a student share his or her result for each drawing. The first drawing should be an irregular hexagon (with two opposing sides of length 1 and four other sides that are longer than 1, and the right drawing should be an irregular pentagon with the base at right angles to its two adjacent sides). Both the hexagon and the pentagon should be bilaterally symmetric.
6. Have the students draw the cross-sections for the pentagonal antiprism. Have a student share his or her result for each drawing. The first drawing should be a regular decagon, and the second drawing should be a parallelogram. The latter is relatively difficult to see.

All eight polygons, properly proportioned, are shown in the below figure.

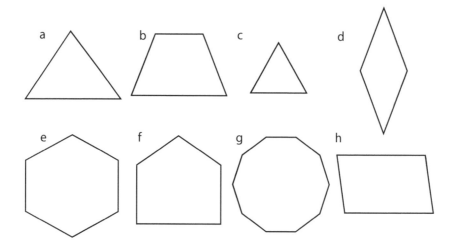

Worksheet 24.1. Cross-sections of polyhedra

Four polyhedra are shown below. All of the edges can be assumed to have length 1. For each one, there are two sets of three points, each of which defines a plane. Draw the polygon that results from slicing the polyhedron by each plane. Then write the name of the polygon below the drawing, being as specific as possible. For example, a right triangle with two sides of the same length should be described as a "right isosceles triangle" rather than simply "triangle" or "isosceles triangle".

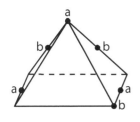 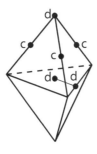 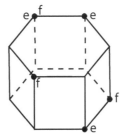 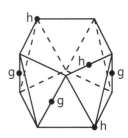

Right square pyramid, plane defined by
    points a (two at midpoints of edges):
    points b (two at the same height):
Triangular dipyramid, plane defined by
    points c (at midpoints of edges):
    points d (narrow line is parallel to an edge):
Right hexagonal prism, plane defined by
    points e:
    points f:
Pentagonal antiprism, plane defined by
    points g (at midpoints of edges):
    points h (one at midpoint of edge):

Chapter 25

Tessellating Other Surfaces

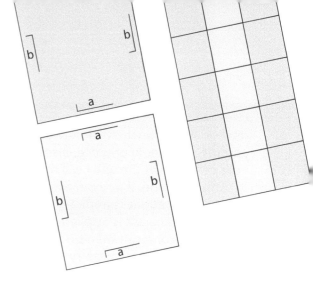

Other surfaces to tessellate

This chapter provides tessellations and nets for three surfaces that are not polyhedra. All of these surfaces are curved, but they only curve in one direction in the vicinity of any point on the surface. This allows them to be made from paper. A surface that curves in two directions, such as a sphere, cannot be constructed by simply bending a sheet of paper.

The first surface is a cylinder. Strictly speaking, a cylinder extends to infinity in either direction along its axis. *An infinite circular* **cylinder** *is the collection of points equidistant from a straight line.* In common usage, the term cylinder is generally taken to mean a right circular cylinder of finite extent, with a circular cap on either end. The cylinder in this chapter is finite in extent, but uncapped. Such

cylinders lend themselves well to tessellation with a repeating design, i.e., one with translational symmetry in the direction perpendicular to the axis. Note that such an application has a practical use in tiling a column. M.C. Escher created tiled columns for a school in Holland using his tessellations.

The second surface is a cone. *A (finite)* **cone** *has both an apex and a base and consists of the locus of all straight-line segments joining the two. The term cone is generally used to refer to a right circular cone, one in which the base is a circle and whose apex lies on a line perpendicular to the plane of the circle and passing through its center.* A model of the surface of a cone, less the base, can be formed from paper by cutting a circle and a radius of that circle. The two open ends are then slipped past one another until the

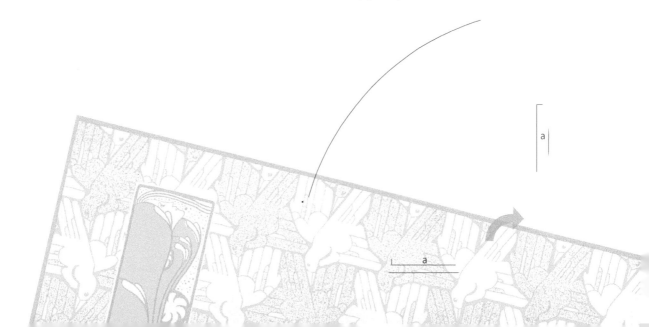

desired angle is achieved between the axis and the surface. In practice, a small circular hole centered at the apex will make it easier to form the cone. If a tessellation possessing n-fold rotational symmetry about a point located at the apex is applied to the paper, a tessellated cone can then be formed with any value of rotational symmetry from 2 through $n - 1$. The owl tessellation in this chapter would have four-fold rotational symmetry in the plane, but has three-fold rotational symmetry when a cone is formed from 3/4 of the circle. If it is further rolled until only half the circle shows, it will have two-fold rotational symmetry. If it is further rolled until only 1/4 of the circle shows, it will no longer have rotational symmetry. (One-fold symmetry doesn't count, as any object when rotated by 360° is unchanged.)

A Möbius strip or Möbius band is an object with one side and one edge. If you start drawing a line along the length of a Möbius strip, you will traverse the entire surface before returning to the starting point, and you will have drawn a line twice as long as the strip. A model of a Möbius strip can be constructed by giving a strip of paper a half twist and joining the ends. This is the sort of model discussed in this chapter, but in order to apply the frog tessellation to both sides, a wide strip is first folded in half to make a narrower strip.

There are several variations on Möbius strips, and they're very entertaining to make and experiment with. For example, if a Möbius strip is cut in half along its length, a single loop with four half twists results. Since the number of half twists is even, the new object is two-sided. If this new strip is cut in half along its length, two independent loops are created, but the two are linked. If a Möbius strip with three half twists is cut in half along its length, a strip that is tied in a trefoil (three crossing) knot results. M.C. Escher depicts a Möbius strip in his print "Möbius strip II (Red Ants)" (1963), and a cut three-half-twist Möbius strip in his print "Möbius strip I" (1961).

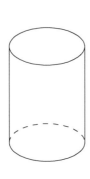

If a cylinder is formed as shown, then there will be four dragons in each row around the circumference. The cylinder can be wound tighter so that there are only three, two, or one dragon(s) in a row.

25a. Cylinder with dragon tessellation

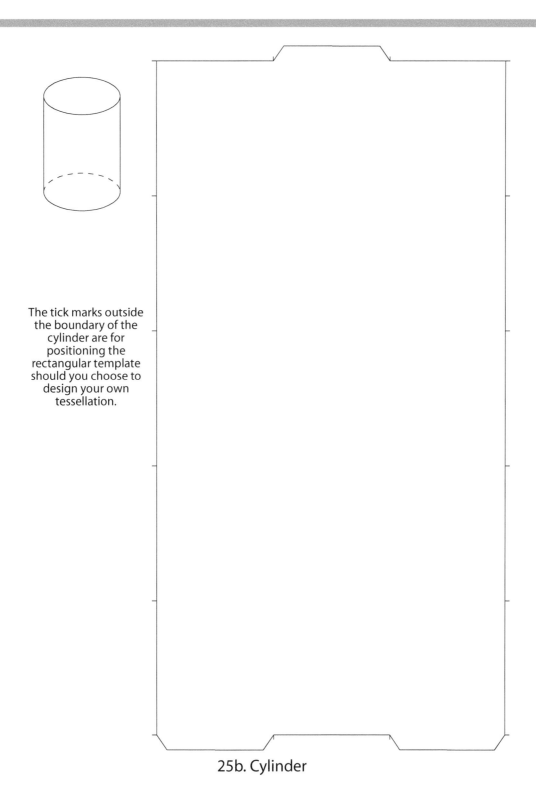

The tick marks outside the boundary of the cylinder are for positioning the rectangular template should you choose to design your own tessellation.

25b. Cylinder

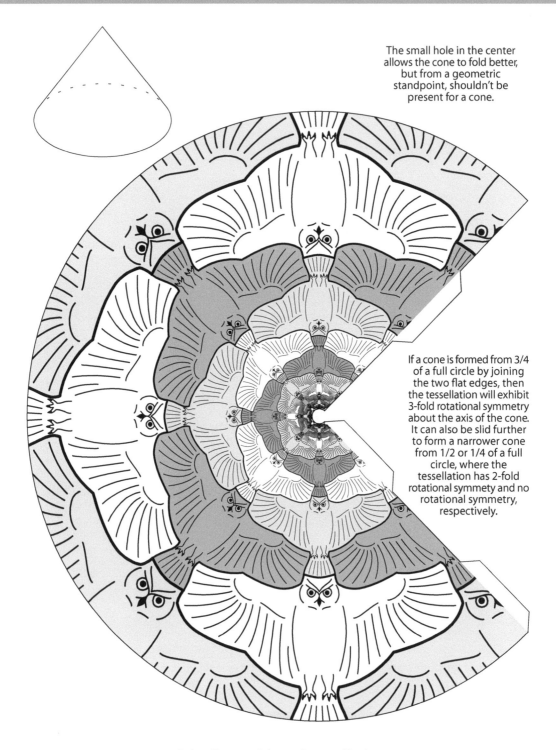

The small hole in the center allows the cone to fold better, but from a geometric standpoint, shouldn't be present for a cone.

If a cone is formed from 3/4 of a full circle by joining the two flat edges, then the tessellation will exhibit 3-fold rotational symmetry about the axis of the cone. It can also be slid further to form a narrower cone from 1/2 or 1/4 of a full circle, where the tessellation has 2-fold rotational symmety and no rotational symmetry, respectively.

26a. Cone with owl tessellation

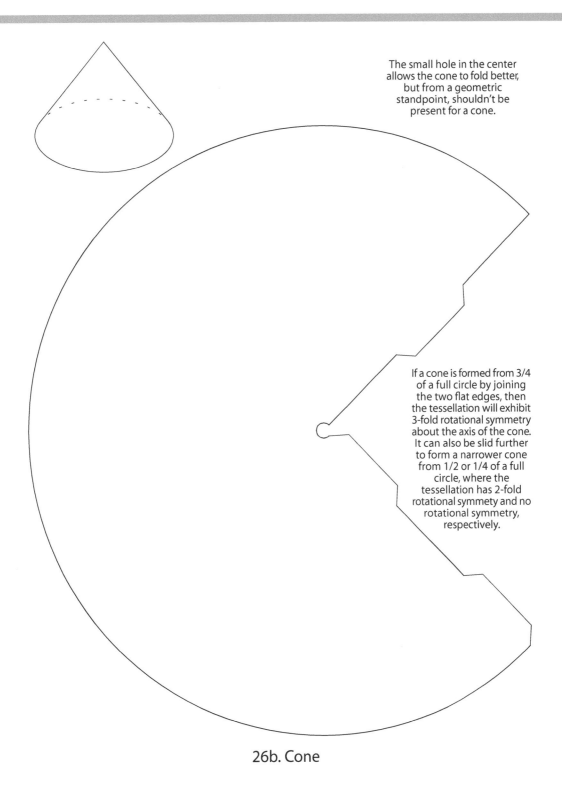

The small hole in the center allows the cone to fold better, but from a geometric standpoint, shouldn't be present for a cone.

If a cone is formed from 3/4 of a full circle by joining the two flat edges, then the tessellation will exhibit 3-fold rotational symmetry about the axis of the cone. It can also be slid further to form a narrower cone from 1/2 or 1/4 of a full circle, where the tessellation has 2-fold rotational symmety and no rotational symmetry, respectively.

26b. Cone

- 1 side
- 1 edge

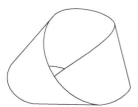

The rectangle should be folded in half along the dotted line and tape or glue used to hold it closed. The strip should then be given a half twist and the tab inserted in the slot between the two folded halves, with tape or glue used to hold the two ends together.

27a. Möbius strip with frog tessellation

- 1 side
- 1 edge

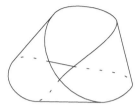

The rectangle should be folded in half along the dotted line and tape or glue used to hold it closed. The strip should then be given a half twist and the tab inserted in the slot between the two folded halves, with tape or glue used to hold the two ends together.

The tick marks outside the boundary of the strip are for positioning the rectangular template should you choose to design your own tessellation.

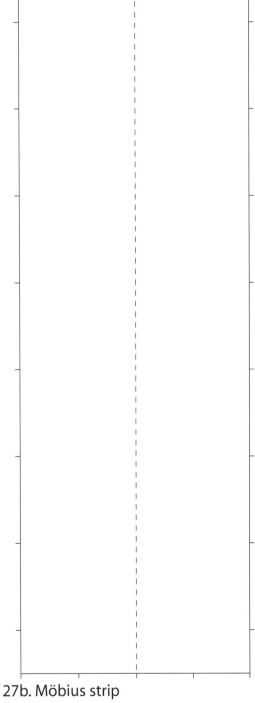

27b. Möbius strip

Tessellation templates for other surfaces

The template shown in Figure 25.1 for the cylinder has the same symmetries as that which was used for the dragon tessellation, but a simpler shape. Note that the two instances of the (b) line segment are reflected about a horizontal axis with respect to one another. This means that two versions of the tile are needed, with one the mirror image of the other. They will fill the cylinder in an alternating rows (the rows that will wrap around the cylinder are vertical in the figure), with all the tiles in the same row being of the same type. The template for the cone can be filled with any combination of tiles, so long as the two (a) line segments are related as shown in Figure 25.2. The Möbius strip template, Figure 25.3, has three independent line segments. Note that the tile will appear in two orientations related by a 180° rotation. They

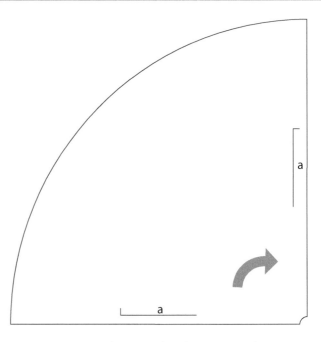

Figure 25.2. Template to aid in designing Escheresque tessellations for a cone.

will fill the strip as indicated, with half tiles at top and bottom. The cylinder and Möbius strip pages include tick marks to aid in locating the tiles.

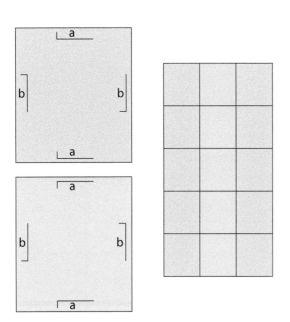

Figure 25.1. Template to aid in designing Escheresque tessellations for a cylinder.

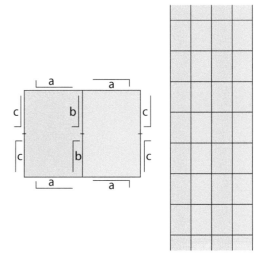

Figure 25.3. Template to aid in designing Escheresque tessellations for a Möbius strip.

Activity 25.1. Surface area and volume of cylinders and cones

Materials: Copies of Worksheet 25.1 for students.

Objective: Apply the surface area and volume formulas for cylinders and cones to solve real-world problems.

Vocabulary: Right circular cylinder, right circular cone.

Specific Common Core State Standards for Mathematics addressed by the activity

Know the formulas for the volumes of cones, cylinders, and spheres and use them to solve real-world and mathematical problems. (8.G.9)

Use volume formulas for cylinders, pyramids, cones, and spheres to solve problems. (G GMD 3)

Activity Sequence

1. Pass out the worksheet.
2. Write the vocabulary terms on the board and discuss the meaning of each one.
3. Have the students write the surface area and volume formulas from memory. Be sure all of the students have the correct formulas.

For the cylinder, $A = 2\pi r^2 + 2\pi rh$ and $V = \pi r^2 h$. For the cone, $A = \pi r^2 + \pi r(r^2 + h^2)^{1/2}$ and $V = \pi r^2 h/3$.

4. Have the students work the first word problem. Have one or more students share their answers.

    28.3 cubic feet and 48 bags (you will need 47.1, but you can't buy a partial bag).

5. Have the students work the second word problem. Have one or more students share their answers.

    This is a little tricky, as the surface area formula cannot be applied directly. The top and bottom of the pillar wouldn't be tiled, and so the relevant area is $2\pi rh$. The answer is 76 sheets. If the top and bottom were left in, it would be 83 sheets.

6. Have the students work the third word problem. Have one or more students share their answers.

    27.2 gallons.

7. Have the students work the fourth word problem. Have one or more students share their answers.

    For each cone, to get the relevant area, you need to subtract the base term (πr^2) from the surface area formula and then double it to account for coating both sides. 648 cubic inches and 21.6 pounds.

Worksheet 25.1. Surface area and volume of cylinders and cones

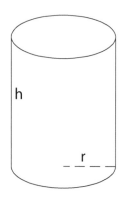
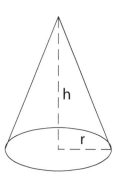

The above figure shows a right circular cylinder and a right circular cone. Next to each one, write the formula for its surface area and volume.

Imagine you have the job of making two tiled, solid concrete pillars for a courtyard. The pillars are to be 8 feet tall and 15 inches in diameter. How many cubic feet of concrete will you need if you want to have just enough for the job? If an 80 pound bag of premixed concrete makes 0.6 cubic feet, how many bags will you need to buy?

If the pillars are to be tiled with small square tiles that come in 1 square foot sheets, how many sheets will you need to buy to tile the two pillars?

Imagine you own an ice cream shop and you expect to sell an average of 1,000 ice cream cones per week. You know that the average amount of ice cream used per cone is 1.5 times the volume of the cone. The cones are 4 inches tall and 2 inches in diameter at the opening. How many gallons of ice cream will you need to order each week, on average? (There are 231 cubic inches in 1 gallon.)

You expect one in five customers to want his or her cone dipped in chocolate, and you know that the chocolate will coat the cone to a thickness of 1/8 inch, inside and out. How many cubic inches of chocolate will you need each week? If the chocolate is sold by the pound, and one pound corresponds to 30 cubic inches, how many pounds will you need to buy?

References

Anderson, S. 2016. http://www.squaring.net.

Araki, Y. 2020. https://www.tessellation.jp/.

Bailey, D. 2020. David Bailey's world of Escher-like tessellations. http://www.tess-elation.co.uk/.

Bareiss S. 2020. http://www.tessellations.org/.

Bellos, A. 2015. The golden ratio has spawned a beautiful new curve: The Harriss spiral. https://www.theguardian.com/science/alexs-adventures-in-numberland/2015/jan/13/golden-ratio-beautiful-new-curve-harriss-spiral.

Bilney, B. 2020. Ozzigami. http://www.ozzigami.com.au/tessellations.html.

Bonner, J. 2017. *Islamic Geometric Patterns*. New York: Springer.

Bool, F.H., J.R. Kist, J.L. Locher, and F. Wierda 1982. *M.C. Escher: His Life and Complete Graphic Work*. New York: Harry N. Abrams.

Bulatov, V. 2010. Conformal models of the hyperbolic geometry. In: *Presented at the Joint Mathematics Meetings*, San Francisco, CA. http://bulatov.org/math/1001/.

Clarke, A. 2019. The poly pages. http://www.recmath.com/PolyPages/index.htm.

Conover, E. 2018. How math helps explain the delicate patterns of dragonfly wings. https://www.sciencenews.org/article/how-math-helps-explain-delicate-patterns-dragonfly-wings.

Conway, J.H., H. Burgiel, and C. Goodman-Strauss 2008. *The Symmetries of Things*. New York: AK Peters / CRC Press.

Coxeter, H.S.M. 1954. Regular honeycombs in hyperbolic space. In: *Proceedings of the International Congress of Mathematicians*, ed. J.C.H. Gerretsen and J. De Groot. Amsterdam: North-Holland Publishing.

Crompton, A. 2020. http://www.cromp.com/pages/home.html.

Darst, R., J. Palagallo, and T. Price 1998. Fractal tilings in the plane. *Mathematics Magazine* 71(1): 12–23.

David, H. 2020. Hop's Gallery. http://www.clowder.net/hop/.

Dunham, D. 2006. More "Circle Limit III" patterns. In: *Proceedings of the Bridges*, ed. R. Sarhangi and J. Sharp, 451–458. London: Tarquin Publications.

El-Said, I. and A. Parman 1976. *Geometric Concepts in Islamic Art*. Palo Alto, CA: Dale Seymour Publications.

Ernst, B. 1976. *The Magic Mirror of M.C. Escher*. New York: Ballantine.

Fathauer, R.W. 2000. *Proceedings of the Bridges*, ed. R. Sarhangi, 285–292. Winfield, KS: Southwestern College.

Fathauer, R.W. 2001. Fractal tilings based on v-shaped prototiles. *Computers and Graphics* 26(4): 635–643.

Fathauer, R.W. 2002. Fractal tilings based on kite- and dart-shaped prototiles. *Computers and Graphics* 25(2): 323–331.

Fathauer, R. 2003. Extending Escher's recognizable-motif tilings to multiple-solution tilings and fractal tilings. In *M.C. Escher's Legacy*, ed. M. Emmer and D. Schattschneider. Berlin: Springer.

Fathauer, R. 2007. Fractal knots created by iterative substitution. In: *Proceedings of the Bridges*, ed. R. Sarhangi and J. Barrallo, 335–342. London: Tarquin Publications.

Fathauer, R. 2009. Fractal tilings based on dissections of polyominoes, polyhexes, and polyiamonds. In: *Homage to a Pied Piper*, ed. E. Pegg Jr., A.H. Schoen, and T. Rodgers. Wellesley, MA: A.K. Peters.

Fathauer, R. 2012. Two-color fractal tilings. In: *Proceedings of the Bridges 2012*, ed. R. Bosch, D. McKenna, and R. Sarhangi, 199–206. Phoenix: Tessellations Publishing.

Fathauer, R. 2014. Some hyperbolic fractal tilings. In: *Proceedings of the Bridges*, ed. E. Torrence, B. Torrence, C. Séquin, and Kristof Fenyvesi, 15–22. Phoenix: Tessellations Publishing.

Fathauer, R. 2014. *Tessellations in Our World*. Phoenix: Tessellations Publishing.

Fathauer, R. 2018. Art and recreational math based on kite-tiling rosettes. In: *Proceedings of the Bridges 2018*, ed. G. Greenfield, G. Hart, and R. Sarhangi, 87–94. Phoenix: Tessellations Publishing.

Fathauer, R. 2019. Fractal diversions. https://www.mathartfun.com/fractaldiversions/index.html.

Fathauer, R. 2020. Robert Fathauer: Art inspired by mathematics and nature. http://robertfathauer.com/.

Frederickson, G. 1997. *Dissections Plane and Fancy*. Cambridge: Cambridge University Press.

Frettlöh, D., E. Harriss, and F. Gähler 2019. Tilings encyclopedia. https://tilings.math.uni-bielefeld.de.

Friedman, E. 2018. https://www2.stetson.edu/~efriedma/mathmagic/1002.html.

Gailiunas, P. 2007. Some monohedral tilings derived from regular polygons. In: *Bridges Donostia*, ed. R. Sarhangi and J. Barrallo, 9–14. London: Tarquin Publications.

Gamepuzzles. 2019. http://www.gamepuzzles.com.

Gardner, M. 1977. Extraordinary nonperiodic tiling that enriches the theory of tiles. *Scientific American* 236: 110–121.

GeoGebra Arts & STEAM. 2019. https://www.facebook.com/groups/GeoGebraSTEAM/.

Gjerde, E. 2009. *Origami Tessellations: Awe-Inspiring Geometric Designs*. Boca Raton, FL: CRC Press.

Golomb, S. 1994. *Polyominoes*. Princeton, NJ: Princeton University Press.

Goodman-Strauss, C. 1999. A small aperiodic set of planar tiles. *European Journal of Combinatorics* 20(5): 375–384.

Grünbaum, B. and G.C. Shephard 1987. *Tilings and Patterns*. New York: W.H. Freeman.

Hart, G. 2019. Colorings. https://www.georgehart.com/virtual-polyhedra/colorings.html.

Hart, G. 2019. Geometric sculpture. https://www.georgehart.com/sculpture/sculpture.html.

Hofstadter, D. 1986. *Metamagical Themas: Questing for the Essence of Mind and Pattern*. New York: Bantam Books.

Jardine, K. 2018. Imperfect convergence. http://gruze.org/tilings/.

Kaplan, C. 2010. Curve evolution schemes for parquet deformations. In: *Proceedings of the Bridges 2010*, ed. G. Hart and R. Sarhangi, 95–102. Phoenix: Tessellations Publishing.

Kaplan, C. 2019. Isohedral – The website of Craig S Kaplan. http://isohedral.ca/blog/.

Kuiper, H. 2020. Computer art by Hans Kuiper. http://web.inter.nl.net/hcc/Hans.Kuiper/.

Landry, K. 2020. http://tessellation.info/en/info/artists/32/Kenneth_Landry/.

Lee, K. 2016. *TesselManiac!* Sandpiper Software and Tessellations.

Leys, J. 2019. Hyperbolic Escher. http://www.josleys.com/show:gallery.php?galid=325.

Lomonaco, S. and L. Kauffman. 2008. Quantum Knots and Mosaics. arXiv:0805.0339v1.

Mandelbrot, B. 1982. *The Fractal Geometry of Nature*. New York: Freeman and Co.

Mann, C. 2004. Heesch's tiling problem. *The American Mathematical Monthly* 111(6): 509–517.

Mann, C., J. McLoud-Mann, and D. Von Derau 2015. Convex pentagons that admit *i*-block transitive tilings. arXiv:1510.01186 [math.MG].

Mikloweit, U. 2019. Polyedergarten. http://www.polyedergarten.de.

Myers, J. 2009. 2-uniform tilings with regular polygons and regular star polygons. https://www.polyomino.org.uk/publications/2009/star-polygon-tiling-2-uniform.pdf.

Myers, J. 2019. https://www.polyomino.org.uk/mathematics/polyform-tiling/.

Nakamura, M. 2020. http://www.tessellation.info/en/info/artists/5/Makoto_Nakamura/.

Nelson, D.R. 1986. Quasicrystals. *Scientific American* 255(2): 42–51.

Nelson, R. 2019. Twitter bot @TilingBot. https://twitter.com/TilingBot.

Osborn, J.A.L. 2020. http://www.ozbird.net/index.html.

Ouyang, P., X. Tang, K. Chung, and T. Yu 2018. Fractal tilings based on successive adjacent substitution rule. *Complexity* 2018: 12.

Ouyang, P. and R. Fathauer 2014. Beautiful math, Part 2: Aesthetic patterns based on fractal tilings. *IEEE Computer Graphics and Applications* 34(1): 68–76.

Pantal, V. 2020. https://twitter.com/panlepan.

Passiouras, S. 2019. Escher tiles. http://www.eschertiles.com/.

Pegg, Jr., E. 2016. Ponting square packing. https://demonstrations.wolfram.com/PontingSquarePacking/.

Pegg, Jr., E. 2019. Square Packing. http://mathpuzzle.com.

Peitgen, H., H. Jürgen, and D. Saupe 1992. *Fractals for the Classroom*. New York: Spring-Verlag.

Ponting, A. 2014. Square packing. http://www.adamponting.com/square-packing/.

Raedschelders, P. 2003. Tilings and other unusual Escher-related prints. In: *M.C. Escher's Legacy*. Berlin: Springer.

Raedschelders, P. 2020. http://tessellation.info/en/info/artists/23/Peter_Raedschelders/.

Rao, M. 2017. Exhaustive search of convex pentagons which tile the plane. arXiv:1708.00274 [math.CO].

Rhoads, G.C. 2005. Planar tilings by polyominoes, polyhexes, and polyiamonds. *Journal of Computational and Applied Mathematics* 174(2): 329–353.

Rice, M. 2020. http://www.tessellation.info/en/info/artists/17/Marjorie_Rice/.

Roelofs, R. 2019. http://www.rinusroelofs.nl/index.html.

Schattschneider, D. 1978. Tiling the plane with congruent pentagons. *Mathematics Magazine* 51(1): 29–44.

Schattscheider, D. 2004. *M.C. Escher: Visions of Symmetry*. New York: Harry N. Abrams.

Schattscheider, D. and W. Walker 1987. *M.C. Escher Kaleidocycles*. Corte Madera, CA: Pomegranate Artbooks.

Scherphuis, J. 2018. Tilings. https://www.jaapsch.net/tilings/index.htm.

Schoen, A. 2020. Finite tilings by rhombs. https://schoengeometry.com/b-fintil.html.

Seymour, D. and J. Britton 1989. *Introduction to Tessellations*. Palo Alto, CA: Dale Seymour Publications.

Sharp, J. 2000. Beyond the golden section - The golden tip of the iceberg. In: *Bridges Conference Proceedings*, ed. R. Sarhangi, 87–98. Winfield, KS: Bridges Conference.

Shechtman, D., I. Blech, D. Gratias, and J. Cahn 1984. Metallic phase with long-range orientational order and no translational symmetry. *Physical Review Letters* 53(20): 1951–1953.

Snels, P. 2019. Tessellations. http://tessellation.info/en.

Socolar, J.W.S. and J.M. Taylor 2011. An aperiodic hexagon tile. *Journal of Combinatorial Theory, Series A* 118(8): 2207–2231.

Specht, E. 2018. http://www.packomania.com.

Stock, D.L. and B.A. Wichmann 2000. Odd spiral tilings. *Mathematics Magazine* 73(3): 239–246.

Taimina, D. 2018. *Crocheting Adventures with Hyperbolic Planes, 2nd ed.* Boca Raton, FL: CRC Press.

Tantrix. 2020. http://www.tantrix.com/english/TantrixHistory.html.

Thomas, S. 2020. Plane liner. http://www.bridgesmathart.org/art-exhibits/bridges06/thomas.html.

Van Deventer, O. 2011. Fractal Jigsaw. https://www.youtube.com/watch?v=7SbOYXQ2x6I.

Vince, A. 1993. Replicating tessellations. *SIAM Journal on Discrete Mathematics* 6(3): 501–521.

Waldman, C.H. 2016. Gnomon is an Island: Whorled polygons. http://old.nationalcurvebank.org/gnomon/Gnomon%20is%20an%20Island.pdf.

Webster, P. 2013. Fractal Islamic geometric patterns based on arrangements of {n/2} stars. In: *Proceedings of the Bridges 2013*, ed. G.W. Hart and R. Sarhangi, 87–94. Phoenix: Tessellations Publishing.

Weeks, J. 2019. KaleidoTile. http://www.geometrygames.org.

Weisstein, E.W. 2018. Various articles. From MathWorld: A wolfram web resource. Tiling http://mathworld.wolfram.com/Tiling.html.

Whittaker, E.J.W. and R.M. Whittaker 1988. Some generalized penrose patterns from projections of n-dimensional lattices. *Acta Crystallographica. Section A* 44(2): 105–112.

Whyte, D. 2020. beesandbombs. https://twitter.com/beesandbombs.

Wichmann, B. and T. Lee 2019. http://www.tilingsearch.org.

Wikipedia. 2018–2020. Various articles. https://www.wikipedia.org.

Glossary of Terms

Admit: If a particular tessellation can be created using a particular prototile, the tile is said to admit the tessellation.

Angle: A measure of the amount of turning necessary to bring a line or other object into coincidence with another.

Antiprism: An antiprism consists of two regular polygons that are rotated with respect to one another so that a band of triangles (equilateral if the antiprism is regular) can connect them.

Aperiodic set of tiles: One that only allows tessellations that do not possess translational symmetry.

Archimedean solid: One for which each face is a regular polygon, but (in contrast to the Platonic solids) not all are of the same type, and for which all vertices are of the same type. Also known as an Archimedean polyhedron, semi-regular solid, or semi-regular polyhedron.

Archimedean spiral: A curve with the equation $r = \theta^{1/n}$, where n is a constant that determines how tightly the spiral is wrapped.

Archimedean tessellation: The regular and semi-regular tessellations together. This term is sometimes used to refer only to the semi-regular tessellations.

Bilateral symmetry: A symmetry in which two halves of an object mirror each other about a center line.

Cantellation: A process in which a rectified tiling is rectified a second time.

Catalan solids: The dual polyhedra to the Archimedean solids.

Centrosymmetric: The property of an object being unchanged by a 180° rotation about its center.

Chamfering: Truncation of edges of a tessellation or a polyhedron.

Chiral polyhedron: One that has distinct left-handed and right-handed versions that are mirror images of one another.

Circle packing: The process of arranging circles that touch tangentially to fill the infinite plane or a figure such as a square.

Coloring (of a tessellation): How colors are assigned to the different tiles in a tessellation.

Composition: The process of merging tiles to form larger tiles.

Concave polygon: One in which at least one interior angle exceeds 180°.

Cone: An object with both an apex and a base and consisting of the locus of all straight-line segments joining the two. The term cone is generally used to refer to a right circular cone, one in which the base is a circle and whose apex lies on a line perpendicular to the plane of the circle and passing through its center.

Congruent: Having the same shape and the same size.

Conway criterion: A set of conditions under which a prototile will tile simply by translation and 180° rotation.

Convex polygon: One in which none of the interior angles exceed 180°.

Convex polyhedron: One for which any line connecting two points on the surface lies in the interior or on the surface of the polyhedron.

Corner (of a tile): A point where two edges of the tile meet.

Corona: A full layer of congruent copies of a tile T surrounding T.

Cylinder: The collection of points equidistant from a straight line. In common usage, the term cylinder is generally taken to mean a right circular cylinder of finite extent, with a circular cap on either end.

Dart: A quadrilateral with two adjacent sides of one length and two other adjacent sides of a second length, and one interior angle greater than 180°.

Decagon: A ten-sided polygon.

Decomposition: The process of dividing tiles into smaller tiles.

Decorated tile: One which has markings in its interior.

Degree (of a vertex): See Valence.

Dihedral angle: The angle between two faces of a polyhedron.

Dihedral tiling: one in which all tiles are congruent to one of two prototiles; mirrors are allowed.

Dipyramid: A dipyramid is formed by two pyramids joined base-to-base.

Dodecagon: A 12-sided polygon.

Double spiral: A spiral that has two points at the perimeter where it can be extended.

Double-armed spiral: A two-ended spiral.

Dual polyhedron: The dual of a polyhedron is the polyhedron obtained by connecting the midpoints of each face with their nearest neighbors.

Dual tessellation: One obtained by connecting the centers of every pair of adjacent tiles in a tessellation.

Edge (of a polyhedron): An edge of a polyhedron is a straight-line segment shared by two adjacent faces.

Edge (of a tile): A line segment joining two corners of the tile.

Edge-to-edge tessellation: One in which adjacent tiles always touch along entire edges.

Elliptic geometry: A geometry in which the angles of tiles meeting at a vertex sum to less than 360°.

Equilateral polygon: One for which each edge is of the same length.

Escheresque tessellation: A tessellation in which the individual tiles are recognizable, real-world motifs.

Escheresque tile: A tile that resembles a real-world motif; in a general sense, the sort of tiles found in the tessellations for which M.C. Escher is famous.

Euclidean plane: The infinite two-dimensional space that obeys the rules of Euclidean geometry, the geometry people are most familiar with.

Exterior angle (of a tile): The angle between two adjacent edges measured outside the tile.

f-tiling: A fractal arrangement of tiles that contains singularities and is bounded in the plane.

Face: A face of a polyhedron is a flat region of polygonal shape.

Fibonacci sequence: An integer sequence in which each number is the sum of the two preceding numbers.

Figure: The part of a scene or drawing that is in the foreground, usually representing a form or forms.

Fractal: An object that exhibits self-similarity on progressively smaller scales.

Fractal tile: A tile with a fractal boundary.

Frieze group: A symmetry group for patterns that repeat in only one direction.

Friezes: Patterns that repeat in only one direction.

Geometric tessellation: A tessellation in which the individual tiles are abstract geometric shapes.

Glide reflection: A motion consisting of a translation and a reflection or mirroring.

Glide reflection symmetry: An object possesses glide reflection symmetry if it can be translated (glide) some amount along a vector, then reflect about that vector and remain unchanged.

Golden number: An irrational number ≈ 1.618 or 0.618 that appears in the diagonal of a regular pentagon, in spiral tilings of squares and golden isosceles triangles, and in aperiodic sets of tiles such as the Penrose tiles. Also known as the golden ratio, golden mean, golden section, sacred cut, and divine proportion.

Grid: See lattice.

Ground: The part of a scene or drawing that is in the background, usually the space around and behind a form or forms.

Heesch number: The maximum number of coronas that can be constructed around a non-tessellating tile.

Heesch type: A classification system for tiles that will tessellate, due to Heinrich Heesch.

Hexagon: A six-sided polygon.

Hexagonal packing: The most efficient possible way to pack the plane with circles of the same size.

Hexagram: A six-pointed star polygon with alternating interior angles of 60° and 240°.

High-symmetry direction: A direction (vantage point) from which an object exhibits mirror or rotational symmetry.

Honeycomb: A tessellation in three or more dimensions, with 3-honeycomb referring to a tessellation of three-dimensional space.

Hyperbolic geometry: A geometry in which the angles of tiles meeting at a vertex sum to greater than 360°.

Infinite: Endless, extending indefinitely.

Interior angle (of a tile): The angle between two adjacent edges measured inside the tile.

Interior details (of a tile): Lines, dots, shading, etc., of the interior of a tile, used to suggest a real-world motif.

Interior (of a tile): The area inside the boundary of a tile.

Isosceles triangle: A triangle in which two of the sides are equal in length. An isosceles right triangle is an isosceles triangle containing a right angle, i.e., one with angles 90°, 45°, and 45°.

Iteration: The repeated application of a series of steps.

Kaleidoscope symmetry: The property of not changing under a set of reflections such as those observed in a kaleidoscope.

Kite: A quadrilateral with two adjacent sides of one length and two other adjacent sides of a second length, and all angles less than 180°.

Knot: A closed curve in three-dimensional Euclidean space that doesn't intersect itself.

Lattice: A periodic arrangements of points in two or more dimensions.

Laves tessellations: The duals of the Archimedean tessellations.

Link: Two or more tangled knots.

Logarithmic spiral: A curve with the equation $r = ae^{b\theta}$, where a and b are constants.

Matching rule: A law specifying how tiles in a tessellation should be fit together.

Mathematical modeling: The use of mathematics to approximate and better understand the physical world.

Mathematical plane: A theoretical flat (two dimensional) surface extending to infinity in all directions.

Metamorphosis (morph for short): A change from one thing to another. A tessellation morph is a tessellation that evolves spatially or temporally.

Mirror symmetry: A three-dimensional (two dimensional) object is said to possess mirror symmetry about a plane (line) if the object can be reflected about that plane (line) and remain unchanged.

Möbius band (or strip): A band with one side and one edge. A model of a Möbius strip can be constructed by giving a strip of paper a half twist and joining the ends.

Monohedral tiling: one in which all tiles are congruent; mirrors are allowed.

Motif: A theme or subject.

Natural tessellation: One found in nature. Natural tessellations are always approximate and finite in extent.

Negative space: A graphic design term that denotes the portion of a design perceived as being in the background.

Net: A net of a polyhedron is a drawing showing the constituent polygons laid out flat in a fashion that would allow them to be folded up to form the polyhedron.

Non-periodic tessellation: One that doesn't repeat; i.e., one not possessing translational symmetry.

Octagon: An eight-sided polygon.

Orbifold notation: It is a more recent symmetry classification system that is popular with mathematicians.

Order (of a vertex): See Valence.

Orientation: The way in which an object is aligned with respect to its surroundings.

Parallelogram: A quadrilateral in which both pairs of opposing sides are parallel. A rhombus is a special case of a parallelogram.

Patch: A finite number of tiles whose union, like a single tile, is a topological disk.

Pattern: A two-dimensional design that (generally) possesses some sort of symmetry.

Pentagon: A five-sided polygon.

Periodic tessellation: One possessing translational symmetry.

Perspective drawing: A drawing approximating on a flat surface a three-dimensional image as seen by the eye.

Phyllotaxis: The arrangement of features in plant growth.

Plane symmetry group: See wallpaper group.

Plastic number: A mathematical constant $\approx$ 1.325 that appears in a spiral tiling of equilateral triangles.

Platonic solid: A convex polyhedron for which each face is a congruent regular polygon and all vertices are of the same type. Also known as a Platonic polyhedron, regular solid, or regular polyhedron.

Polyform: A polygon created by joining simple polygons in an edge-to-edge fashion.

Polygon: A closed plane figure made up of straight-line segments.

Polygonal tile: One in which the edges are all straight lines.

Polyhedron: A geometric solid, the faces of which are polygons.

Polyhex: A shape made up of regular hexagons joined in an edge-to-edge fashion.

Polyiamond: A shape made up of equilateral triangles joined in an edge-to-edge fashion.

Polyomino: A shape made up of squares joined in an edge-to-edge fashion.

Positive space: A graphic design term that denotes the portion of a design perceived as being in the foreground.

Primitive unit cell: The smallest portion of a pattern that will generate the full pattern by any combination of translations, rotations, and glide reflections.

Prism: A (regular right) prism has two congruent regular polygon faces separated by a band of squares.

Prototile: A tile to which many or all other tiles in a tessellation are similar (have the same shape).

Pyramid: A (regular right) pyramid consists of a regular polygon (the "base") and triangles meeting at a common vertex (the "apex") located at a point that lies along a line including the centroid of the base and perpendicular to the base.

Pythagorean tiling: The common periodic tessellation with two sizes of squares.

Quadrilateral: A four-sided polygon.

Realistic drawing: One in which objects are drawn to resemble real-world objects as nearly as possible.

Rectangle: A quadrilateral in which every angle is 90°. A square is a special case of a rectangle.

Rectification: A process in which the midpoints of each edge are connected to their nearest neighbors, followed by erasure of the original edges.

Reflection: A transformation in which each point is replaced by a point at the opposite position with respect to a line or plane, as if by a mirror.

Regular polygon: One for which each edge is of the same length and each interior angle has the same measure.

Regular tessellation: An edge-to-edge tessellation by congruent regular polygons.

Rhombic dodecahedron: A solid with 12 faces that are rhombi.

Rhombic triacontahedron: A solid with 30 faces that are rhombi.

Rhombus: A quadrilateral in which every side is of the same length.

Right circular cone: A (finite) cone has both an apex and a base and consists of the locus of all straight-line segments joining the two. In a right circular cone, the base is a circle and the apex lies on a line perpendicular to the plane of the circle and passing through its center.

Right circular cylinder: A solid consisting of two congruent circular faces with their perimeters joined by a band such that the band meets the circles at right angles.

Right triangle: One that contains a 90° angle.

Rosette: A design or an object that is flowerlike or possesses a single point of rotational symmetry.

Rotation: A turning about a point (in two dimensions) or a line (in three dimensions).

Rotational symmetry: A three-dimensional (two dimensional) object is said to possess n-fold rotational symmetry about a line (point) if the object can be rotated by 1/n of a full revolution about that line (point) and remain unchanged.

Schläfli symbol: A pair of integers in the form {n, k} that fully specify a regular tessellation in terms of the number of sides of n of each polygon and the number of polygons k meeting at each vertex.

Self-replicating tile (reptile): A tile that can be tiled by smaller copies of itself.

Semi-regular tessellations: An edge-to-edge tessellation for which every tile is a regular polygon, each vertex is of the same type, and there is more than one type of regular polygon.

Similar: Having the same shape, but not necessarily the same size.

Singular point (or singularity): A point near which tiles become infinitesimally small, and the density of tiles blows up.

Skew: To distort an object such that all points on one axis remain fixed, and other points are shifted parallel to that axis by a distance proportional to their perpendicular distance from the axis. Another word for skew is shear.

Source material: Items, such as photographs and illustrations, used as reference material to assist in the creation of a drawing, etc.

Space-filling polyhedron: A polyhedron that will tile (fill) three-dimensional space without gaps or overlaps.

Spherical geometry: A type of elliptic geometry where everything occurs on the surface of a sphere.

Spherical polyhedron: A projection of a polyhedron onto the surface of a sphere. Spherical polyhedra are a type of spherical tessellation.

Spherical tessellation: A tessellation on the surface of a sphere; i.e., a collection of shapes that cover the surface of a sphere without gaps or overlaps.

Spiral tessellation: A tessellation that emanates from a finite region that the tiles move farther away from as they approximately follow a spiral curve.

Squaring the square: The mathematical recreation of tiling a square with smaller squares with edge lengths that are all integer.

Star polygon: One in which the interior angles alternate between being greater than 180° and less than <180°; a star-like figure without internal lines.

Stella octangula: A polyhedron that is the only stellation of the octahedron. It can be seen as two interpenetrating tetrahedra.

Stellation: The extension of the edges of a regular polygon to form a star-like figure or the extension of the faces of a polyhedron to form a polyhedron that is star-like.

Stylized drawing: One in which objects are intentionally drawn in a manner or style that doesn't conform to the real world.

Substitution tiling: A tiling that is constructed by iterative substitution of tiles according to a set of rules.

Symmetry: The property of not changing under some transformation.

Symmetry group: A classification of a two-dimensional pattern based on the symmetries in the pattern.

Tessellation: A collection of shapes (tiles) that fit together without gaps or overlaps to cover the infinite mathematical plane.

Tessera: A small square or cube of stone or glass used for making mosaics.

Tile: A set whose boundary is a single simple closed curve; an individual shape in a tessellation or tiling.

Tiling: A collection of shapes (tiles) that fit together without gaps or overlap to cover the infinite mathematical plane.

Transformation: For the purposes of this book, a transformation is a manipulation of a two- or three-dimensional shape that changes its position through rotation, translation, and/or glide reflection.

Translation: A movement in some direction by some amount with rotating or reflecting.

Translational symmetry: An object possesses translational symmetry if it can be moved (translated) by some amount in some direction and remain unchanged.

Translation criterion: a set of conditions under which a prototile will tile by translation.

Translation unit cell: The smallest portion of a pattern that will generate the full pattern by translation only.

Trapezoid: A quadrilateral in which one pair of opposing sides are parallel.

Triangulation: A process in which the polygons in a tessellation are converted into triangles by adding edges.

Truncation: Cutting off. A truncated tessellation is one in which each vertex is cut and replaced by a new tile. A truncated polyhedron is one that can be formed by cutting off portions of another polyhedron.

Uniform tiling: One in which the tiles are regular polygons and the vertices are all of the same type.

Unit cell: The smallest group of tiles in a tessellation that may be repeatedly copied and used to generate the entire tessellation through translation only.

Utilitarian: Of practical use.

Valence: The number of tiles meeting at a vertex in a tessellation, or the number of faces meeting at a vertex in a polyhedron.

Vector: An arrow indicating both distance and direction.

Vertex: A point at which three or more tiles meet in a tessellation (2D) or a point at which three or more faces meet in a polyhedron (3D).

Viewpoint: Point of view; position from which something is looked at.

Voderberg tilings: A spiral tiling in which each tile is congruent, and in which the tiles have the property that two copies can completely surround one or two copies of the tile.

Voronoi tessellation: One created by drawing perpendicular lines midway between the lines joining neighboring points in a collection of points and using those lines to define tiles.

Wallpaper group: The collection of all symmetries possessed by a pattern that repeats in two directions; also referred to as plane symmetry group.

Index